THE HEDGEHOG, THE FOX, AND THE MAGISTER'S POX

THE HEDGEHOG,

MENDING THE GAP BETWEEN
SCIENCE AND THE HUMANITIES

THE FOX, AND THE MAGISTER'S POX

STEPHEN JAY GOULD

Harmony Books
NEW YORK

Published by Harmony Books, New York, New York.
Member of the Crown Publishing Group, a division of Random House, Inc.
www.randomhouse.com

HARMONY BOOKS is a registered trademark and the Harmony Books colophon
is a trademark of Random House, Inc.

Printed in the United States of America

Design by Lynne Amft

Library of Congress Cataloging-in-Publication Data
Gould, Stephen Jay.
The hedgehog, the fox, and the magister's pox : mending the gap
between science and the humanities / Stephen Jay Gould.—1st ed.
Includes index.
1. Science—Social aspects.   2. Science and state.   I. Title.
Q175.55 .G68   2003
303.48′3—dc21   2002007807

ISBN 0-609-60140-7

10 9 8 7 6 5 4 3 2 1

First Edition

For the American Association for the Advancement of Science (AAAS), a truly exemplary organization, serving so well as the "official" voice of professional science here and elsewhere. With thanks for allowing me to serve as their president and then chairman of the board during the millennial transition of 1999–2001. This book began as my presidential address in 2000. The address is then traditionally published in *Science* magazine, the association's official organ, and America's best general journal for scientific professionals. And with apologies to the editor, Don Kennedy, one of the finest people I have ever known in the world of intellectuals. I promised to follow the tradition, but failed because I soon realized that I needed to write at greater length than I could ever ask you to publish. Thus, I now present the printed version of my presidential address (obviously greatly expanded, for I did not filibuster on your podium), and I dedicate this book to AAAS. It was truly a pleasure and privilege to serve—a line often intoned as the boilerplate of a meaningless cliché, but stated, this time in a heartfelt manner, by a quintessential non-joiner who enjoyed the work and truly gained more than he could ever give.

Contents

A NOTE TO THE READER

The Hedgehog, the Fox, and the Magister's Pox is the last of the seven books that Stephen Jay Gould contracted to write for Harmony Books. It was my privilege to be his editor, and it is an honor to have been asked to write a brief note for this signal volume.

Several years ago, I received a catalog for an auction of decommissioned museum pieces. Being especially interested in amber and fossils, I flipped through the catalog pages, marveling at the amazing variety of pieces that included triplets of trilobites and other vanished creatures frozen in tumbling poses like puppies in strange prehistoric attitudes of play. In the middle of the catalog, I came upon a letter penned by Charles Darwin to an unknown correspondent. I have an enormous admiration of the great man, instilled in me by dedicated science teachers and by years of reading Gould's essays and books, but I had never imagined that such a relic could be owned and contemplated by a layperson. I had to have it.

Months later, having happily triumphed in the auction, I received the letter, framed with glass on both sides to enable a full view. I was excited, yet, as I tried to read it, was immediately dismayed to find that I could barely make out two words in succession. Darwin's penmanship was atrocious. After poring over the letter and drawing up a map of those words that I felt sure I'd interpreted correctly, juxtaposed with many guesses and question marks and not a few blanks, I still had very little sense of the meaning of my prized possession.

At that time, I was working with Steve on his book *Rocks of Ages*. I mentioned my acquisition as well as my frustration to him; he was interested to see the letter and generously agreed to try to help me figure it out. He told me that Darwin was renowned for his illegible script and that he was one of the few people who had ever had the talent for deciphering it. This he did for me, writing the missing words on my map in his own (somewhat) clearer hand, along with a couple of notations, reproduced here in brackets.

Down Bromley Kent
Ap 30/81

My Dear Sir
    I must send you a line to thank you for your "Ice & Water" which I also saw with interest very much [This sentence doesn't make much sense, so I may well be wrong here. I think everything else is pretty surely right. The "also saw" words are particularly badly scrawled]; though I believe we split a little about solid glacier ice and icebergs.—Thanks, also for extract out of newspaper about Rooks and Crows— [Leslie: This must be right. Darwin was interested in the taxonomy and names of these birds] I wish I dared trust it. I see in cutting pages half-an-hour ago, that you fulminate against the skepticism of scientific men.—You would not fulminate quite so much, if you had had my many wild-goose chases after facts stated by men not trained to scientific accuracy. I often vow to myself that I will utterly disregard every statement made by any man who has not shown the world he can observe accurately. I wish I had space to tell you a curious History, which I was fool enough to investigate on almost universal testimony of Beans growing this year upside down. —I firmly believe that accuracy is the most difficult quality to acquire.—I did not, however, intend to say all this.—I very thoroughly enjoyed my half-hour's talk at your pleasant House. —I have been corresponding with Mr. Davidson on the genealogy of Brachiopods; and he will someday, I believe, discuss subject as we wish. He has seen Galton's talk of species grouped like a tree. Mr. D is not at all a full believer in great changes of species which will make his work all the more valuable.—I have also written to Mr. Jamison, urging him to take up Glen Roy. My dear Sir Your very sincerely C Darwin

As Steve told me, it's a "nice letter, a good letter, an interesting letter," although not an important letter. "But, it was written only a couple days *before* an important letter." He was happy to have had the opportunity to read it. Along with his translation, he sent me a photocopy of a catalog page and wrote, "This is the author Darwin refers to in your letter—Davidson/ Brachiopods. Pretty pricey and a classic work." Indeed, T. Davidson's *British Fossil Brachiopods*, with 234 plates, six volumes in seven, cloth, 1851–86, was priced at 490 pounds sterling about four years ago, confirming another Darwinian prediction.

Steve's death still seems impossible. He was at the fulcrum of so much activity. For almost a decade, I'd been speaking with him and his literary agent, Kay McCauley, about a book he planned to write, centered on the intense, early twentieth-century correspondence, which he owned, of two paleontologists. He also planned to write about realized geniuses unrecognized in their time. But these are the unrealized books of a recognized genius. It is a tragedy for readers that we have lost Stephen Jay Gould, the great writer, the irreplaceable teacher, the pioneering researcher and creative thinker, the champion and defender of scientific education. Even given the wealth of brilliant work he has left us, his death is made worse by our loss of his unwritten thoughts, his unrecorded insights, the connections that only he could make, but had yet to make. To borrow a verse, "Gould thou shouldst be living at this hour, The world hath need of thee . . . "

Yet Stephen Jay Gould has indisputably left behind many great treasures, one of the last of which you hold in your hands. *The Hedgehog, the Fox, and the Magister's Pox* is of particular interest because it is an original book, not a collection of his previously published essays from *Natural History* magazine, and his last book on natural history. *Triumph and Tragedy in Mudville,* his baseball memoir, also remains to be published. Steve also left his biological family, his many friends, an extended family of students, colleagues, and readers whom he inspired, his intellectual line, which, like the description of evolution in *Full House,* will prove to be "a copiously branching bush with innumerable present outcomes, not a highway or ladder with one summit."

Steve's brilliant and challenging works, his amazing energy and insights, and his urge to examine that which had yet to be explained will continue to inspire readers, students, and other scientists for generations to come. In his dedication for his book *Rocks of Ages,* Steve wrote to his two sons that they "will have to hold on beyond their father's watch." We, his readers, have to hold on, too, as Steve writes in his preface to this book, to our ethical principles, our commitment to the great experiment of democracy, and our commitment to many paths of intellectual inquiry in the sciences and the humanities "that make our lives so varied, so irreducibly, and so fascinatingly, complex."

In Stephen Jay Gould's books, his voice and purpose are beautifully preserved, clearly visible, literary amber. With my Darwin letter, I also acquired several pretty little pieces of amber in which float flower fragments, a flower bud, and a tiny complete flower and leaves. Whenever I look at these bits, I flash to exchanges I had with Steve that revealed his extraordinary mind, his generosity as a teacher, his joy in discovery and knowledge, and his Darwin-like scrupulousness in observation, writing, and research. I long to hear a disquisition by Steve on these botanicals, just as he translated my Darwin letter and put it into historical context. But it will be up to me to investigate the amber without his guidance and to ferret out the important letter related to my letter, and the evolutionary and humanistic forces they reveal. And it is up to you, the reader, to investigate these final writings of Steve's without his last ministrations and corrections. For Steve died before he could proof the manuscript for this book, before he could double-check his facts and figures, before he could correct the page proofs. So if there are any errors floating in the text, think of them as bits in amber left for you to decipher and puzzle over, and perhaps correct, left by one of the greatest forces in scientific thought and writing with whom we have been privileged to live and from whom we have been privileged to learn for a while, and, through his books, forever.

Leslie Meredith
Senior Editor
November 2002

Introducing the Protagonists

I PREFER THE MORE EUPHONIOUS RUSSIAN BEGINNING FOR FAIRY TALES to our equivalent "once upon a time"—*zhili byli* (or, literally, "lived, was"). Thus I begin this convoluted tale of initial discord and potential concord: "*Zhili byli* the fox and the hedgehog." In his *Historia animalium* of 1551, Konrad Gesner, the great Swiss scholar of nearly everything, drew the initial and "official" pictures of these creatures in the first great compendium of the animal kingdom published in Gutenberg's era. Gesner's fox embodies the deceit and cunning traditionally associated with this important symbol of our culture—poised on his haunches, ready for anything, front legs straight and extended, hindquarters set to spring, ears cocked, and hair erect down the full line of his back. Above all, his face grins enigmatically and throughout, from the erect eyelashes to the long smirk, ending at the tapered nose with widespread whiskers—all seeming to say, "Watch me now, and then tell me if you've ever seen anything even half so clever."

The hedgehog, by contrast, is long and low, all exposed and nothing hidden. Spines cover the entire upper surface of his body; and his small feet neatly fit under this protective mat above. The face, to me, seems simply placid: neither dumb nor disengaged but rather serenely confident in a quiet, yet fully engaged manner.

I suspect that Gesner drew these two animals to emphasize these feelings and associations in a direct and purposeful way. For the *Historia animalium* of 1551 is not a scientific encyclopedia in the modern sense of presenting factual information about natural objects, but rather a Renaissance compendium for everything ever said or reported by human observers or moralists about animals and their meanings, with emphasis on the classical authors of Greece and Rome (seen by the Renaissance as the embodiment of obtainable wisdom in its highest form), and with factual truth and falsity as, at best, a minor criterion for emphasis. Each entry includes empirical information, fables, human uses, and stories and lists of proverbs featuring the creature in question.

The fox and the hedgehog not only embodied their separate and well-known symbols of cunning versus persistence. They had also, ever since the seventh century B.C., been explicitly linked in one of the most widely known proverbs about animals, an enigmatic saying that achieved renewed life in the twentieth century. Gesner clearly drew his fox and hedgehog in their roles as protagonists in this great and somewhat mysterious motto.

In Gesner's time, and ever since for that matter, any scholar in search of a proverb would turn immediately to the standard source, the Bartlett's beyond compare for this form of quotation: the *Adagia* (adages, or proverbs) compiled, and first published in 1500, by the greatest intellectual of the Renaissance, Erasmus of Rotterdam (1466–1536). Gesner, of course, directly used and credited Erasmus's exhaustive discussion of the linking proverb in both his articles, *De Vulpe* (on the fox) and *De Echino* (on the hedgehog) of his 1551 founding treatise.

This somewhat mysterious proverb derives from a shadowy source, Archilochus, the seventh-century B.C. Greek soldier-poet sometimes considered the greatest lyricist after Homer, but known only from fragments and secondary quotations, and not from any extensive writings or biographical data. Erasmus cites, in his universalized Latin, the Archilochian contrast of fox and hedgehog: *Multa novit vulpes, verum echinus unum magnum* (or, roughly, "The fox devises many strategies; the hedgehog knows one great and effective strategy").

I use this well-trodden, if enigmatic, image in two important ways (and in the book's title as well) to exemplify my concept of the proper relationship between the sciences and humanities. I could not agree more with the vital sentiment expressed by my colleague E. O. Wilson (although Part III of this book will also explain my reasons for rejecting his favored path toward our common goal): "The greatest enterprise of the mind has always been and always will be the attempted linkage of the sciences and the humanities" (from his book *Consilience,* Knopf, 1998, page 8). I use Archilochus's old image, and Erasmus's extensive exegesis, to underscore my own recommendations for a fruitful union of these two great ways of knowing. But my comparison will not be based on the most straightforward or simpleminded comparison. That is, I emphatically *do not claim* that one of the two great ways (either science or the humanities) works like the fox, and the other like the hedgehog.

Of my two actual usages, the first is, I confess, entirely idiosyncratic, fully concrete, and almost as enigmatic as the proverb itself. That is, I shall refer, in a crucial argument, to the specific citation of Erasmus's explication of Archilochus's motto as preserved in one particular copy of Gesner's 1551 book. Moreover, although I regale you with foxes and hedgehogs in this introduction, this first usage will now disappear completely from the text until the very last pages, when I cite (and picture) this passage to make a closing general point with specific empirical oomph. As to the equally mysterious Magister who shares titular space with the fox and hedgehog, he will make a short intermediary appearance (in chapter 4) and then also withdraw until his meeting with the two animals on the closing pages.

But my second usage pervades the book, although I try to keep explicit reminders to a bearable minimum (an effort demanding great forbearance, and courting probable failure in any case, from such a didactic character as yours truly). This second employment also sticks closely to the metaphorical meanings that have been grafted upon Archilochus's image throughout history, especially since Erasmus's scholarly exegesis. This usage became central to twentieth-century literary commentary when Isaiah Berlin—my personal intellectual hero, and a wonderful man who befriended me when I was a shy, beginning, absolute nobody—invoked the pairing of fox and hedgehog to contrast the styles and attitudes of several famous Russian writers. Ever since then, scholars have played a common game in designating their favorite (or anathematized) literati either as hedgehogs for their tenacity in sticking to one style or advocating one key idea, or as foxes for their ability to move again and again, like Picasso, from one excellence to an entirely different mode and meaning of expression. The game maintains sharp edges because these attributions have been made both descriptively and proscriptively, and people of goodwill (and bad will too, for that matter) can argue forever about either and both. (I must also confess that I named one of my books of essays *An Urchin in the Storm,* to designate my own stubborn invocation of Darwinian evolution as a subject to fit nearly any context or controversy. Hedgehogs, to Englishmen, are urchins.)

Erasmus (and I am quoting from my 1599 edition of his *Adagia*) begins with the usual and obvious reasons for Archilochus's famous contrast. When pursued by hunters, the fox figures out a new and sneaky way to escape each time: *Nam vulpes multijugis dolis se tuetur adversus venatores* (for the fox defends itself against the hunters by using many different guiles). The hedgehog, on the other hand, tries to keep out of harm's way, but will use its one great trick if overtaken by the hunters' dogs: the animal rolls up into a ball, with its small head and feet, and its soft underbelly, tucked up neatly and completely within the enclosing surface of spines. The dogs can do what they wish: poke the animal, roll it about, or even try to bite, but all to no avail (or to painful injury); for the dogs cannot capture such a passive and prickly ball, and must ultimately leave the animal alone, eventually (when the danger has passed) to unroll and calmly walk away. Erasmus writes: *Echinus unica duntaxat arte tutus est adversus canum morsus, siquidem spinis suis semet involuit in pilae speciem, ut nulla ex parte morsu, prendi queat.* (The hedgehog only has one technique to keep itself safe against the dogs' bite, since it rolls itself up, spines outward, into a kind of ball, so that it cannot be captured by biting.)

Later on in this exegesis, Erasmus even adds an old tale of intensification, delicately mentioning only the outline of the story, and referring his readers to the original sources if they wish to know more. If this one great trick seems to be failing, the hedgehog often ups the same basic ante by letting fly a stream of urine, covering the spines, and weakening them to the point of excision. But how can this dramatic form of self-imposed haircut help the creature? Erasmus goes no further, but when we turn to Pliny and Aelianus (the two classical sources cited by Erasmus), we learn what a tough and determined little bastard this apparently timid creature can be. The ultimate urine trick, we are told, can work in three possible ways. First, with the spines excised, the animal can often slither away unnoticed. Second, the urine smells so bad that the dogs or human hunters may simply lose interest and beat a quick retreat. Third, if all else fails, and the hunters take him anyway, at least the hedgehog can enjoy his last laugh in death, for his haircut has rendered him useless to his captors (who, in a fourth potential utility, might also abandon him in frustration by recognizing this outcome in advance)—for the main attraction of the hedgehog to humans lies in the value of his hide, but only with spines intact, as a natural brush.

The power and attraction of Archilochus's image lies, rather obviously, in its two levels of metaphorical meaning for human contrasts. The first speaks of psychological styles, often applied for quite practical goals. Scramble or persist. Foxes owe their survival to easy flexibility and skill in reinvention, to an uncanny knack for recognizing (early on, while the getting remains good) that a chosen path will not bear fruit, and that either a different route must be quickly found, or a new game entered altogether. Hedgehogs, on the other hand, survive by knowing exactly what they want, and by staying the chosen course with unswerving persistence, through all calumny and trouble, until the less committed opponents eventually drop away, leaving the only righteous path unencumbered for a walk to victory.

The second, of course, speaks to favored styles of intellectual practice. Diversify and color, or intensify and cover. Foxes (the great ones, not the shallow or showy grazers) owe their reputation to a light (but truly enlightening) spread of real genius across many fields of study, applying their varied skills to introduce a key and novel fruit for other scholars to gather and improve in a particular orchard, and then moving on to sow some new seeds in a thoroughly different kind of field. Hedgehogs (the great ones, not the pedants) locate one vitally important mine, where their particular and truly special gifts cannot be matched. They then stay at the site all their lives, digging deeper (because no one else can) into richer and richer stores from a mother lode whose full generosity has never before been so well recognized or exploited.

I use the fox and hedgehog as my model for how the sciences and humanities should interact because I believe that neither pure strategy can work, but that a fruitful union of these seemingly polar opposites can, with goodwill and significant self-restraint on both sides, be conjoined into a diverse but common enterprise of unity and power. The way of the hedgehog cannot suffice because the sciences and humanities, by the basic logics of their disparate enterprises, do different things, each equally essential to human wholeness. We need this wholeness above all, but cannot achieve the goal by shearing off the legitimate differences (I shall critique Wilson's notion of consilience on this basis) that make our lives so varied, so irreducibly, and so fascinatingly, complex. But if we lose sight of the one overarching goal—the hedgehog's insight—underneath the legitimately different concerns and approaches of these two great ways, then we are truly defeated, and the dogs of war will disembowel our underbellies and win.

But the way of the fox cannot prevail either, because too great a flexibility may lead to survival of no enduring value—mere persistence with no moral or intellectual core intact. What triumph can an ultimate chameleon claim if he gains not even the world, but only his basic continuity, at the price of his soul? Fortunately, and in the most parochial American sense, we know a model of long persistence and proven utility for the virtues in fruitful union of apparent opposites. This model has sustained us through the worst fires of challenge (both voluntary self-immolation from 1861 to 1865, and attempted external prevention at several times, beginning with the first battles of 1775).

We have even embodied this ideal in our national motto, *e pluribus unum,* "one from many." If the different skills and wondrous flexibilities of the fox can be combined with the clear vision and stubbornly singleminded goal of the hedgehog, then a star-spangled banner can protect a great expanse of maximal diversity because all the fox's skills now finally congeal to realize the hedgehog's great vision. Never before in human history has the experiment of democracy been tried across such a vast range of geographies, climates, ecologies, economies, languages, ethnicities, and capabilities. Lord knows we have suffered our troubles, and imposed horrendous and enduring persecutions upon sectors of the enterprise, thus sullying the great goal in the most shameful way imaginable. Yet, on balance, and by comparison to all other efforts of similar scale in human history, the experiment has worked, and has been showing substantial improvement in the course and memories of my lifetime at least.

I offer the same basic prescription for peace, and mutual growth in strength, of the sciences and humanities. These two great endeavors of our soul and intellect work in different ways and cannot be morphed into one simple coherence, so the fox must have his day. But the two enterprises can lead us onward together, ineluctably yoked if we wish to maintain any hope for arrival at all, toward the common goal of human wisdom, achieved through the union of natural knowledge and creative art, two different but nonconflicting truths that, on this planet at least, only human beings can forge and nurture.

But I learned one other important lesson from reading Erasmus's commentary, and by considering the deeper meaning of Gesner's pictures.

Erasmus does, following the literal lead of Archilochus's minimality, depict the styles of the fox and hedgehog as simply different, with each strategy effective in its own way, and expressing one end of a full continuum. But Erasmus clearly favors the hedgehog in one crucial sense: foxes generally do very well indeed, but when the chips go down in extremis, look inside yourself, and follow the singular way that emerges from the heart and soul of your ineluctable being and construction, whatever the natural limits—for nothing beats an unswerving moral compass in moments of greatest peril.

Erasmus, after praising the many wiles of the fox (as quoted above), then adds *et tamen haud raro capitur*—"yet, nonetheless, it is captured not rarely." The hedgehog, on the other hand, almost always emerges unscathed, a bit stressed and put-upon, perhaps, but ultimately safe nonetheless. And thus intellectuals of all stripes and tendencies must maintain this central integrity of no compromise to fashion or (far worse) to the blandishments of evil in temporary power. We have always been, and will always be, a minority. But if we roll with the punches, maintain the guts of our inner integrity, and keep our prickles high, we can't lose—for the pen, abetted by some modern modes of dispersal, really is mightier.

Finally, I don't mean to despise or dishonor the fox, and neither does Erasmus, despite his clear zinger, quoted just above, against this ultimate symbol of wiliness. For Erasmus ends his long and scholarly commentary with two stories about dialogues between the fox and another brother carnivore. The first tale of the fox and cat simply extends Erasmus's earlier point about the hedgehog's edge in episodes of greatest pith and moment. The two animals meet and begin to argue about better ways to elude packs of hunting dogs. The fox brags about his enormous bag of tricks, while the cat describes his single effective way. Then, right in the midst of this abstract discussion, the two creatures must face an unexpected and ultimately practical test: "Suddenly, amidst the dispute, they hear the voices of the dog pack. The cat immediately leaps up into the highest tree, but the fox, meanwhile, is surrounded and captured by the crowd of dogs." *Praestabilius esse nonnunquam unicum habere consilium* (perhaps it is better to have one way of wisdom), Erasmus adds, *id sit verum et efficax* (provided that it be true and effective).

But the second tale of the fox and panther saves our maligned character and shows the inner beauty of his flexibility, as illustrated by his avoidance of mere gaudy show for true dexterity of mind. Erasmus writes:

Cum aliquando pardus vulpem pre se contemneret, quod ipse pellem haberet omnigenus colorum maculis variegatem, respondit vulpes, sibi decoris in animo esse, quod ille esset in cute.

"When the panther disparages the fox by comparison to himself, because his [the panther's] skin is so beautifully variegated with so many colored spots of all kinds, the fox responds that it is better to be so decorated in the mind than upon the skin."

And so I say to the sciences (where I reside with such lifelong pride and satisfaction) and to the humanities (whose enduring technique of exegesis from printed classical sources I try, in my own conceit, to utilize as the primary mode of analysis in this book): what a power we could forge together if we could all pledge to honor both of our truly different and equally necessary ways, and then join them in full respect, in the service of a common goal as expressed in old Plato's definition of art as intelligent human modification and wondrous ornamentation, based on true veneration of nature's reality. For then, as the Persian poet said:

Oh wilderness were Paradise enow.

Then wilderness (nature's unvarnished tangle of wonders) would become a paradise (literally, a cultivated garden of human delight).

The goal could not be greater or more noble, but the tensions are old and deep, however falsely construed from the start, and stirred up by small minds ever since. Thus the union of the fox and hedgehog can certainly be accomplished, and would surely yield, as progeny, a many-splendored thing called love and learning, creativity and knowledge. But we had best proceed, in this hybridization, by the resolution of a bad old joke about an animal not closely related to the hedgehog, but functionally equivalent in the primary manner of this discussion. How, using more decorous language than the joke enjoins, can two porcupines copulate? The answer, of course, is "carefully."

I

THE RITE AND
RIGHTS OF A
SEPARATING SPRING

1

Newton's Light

THE EPITAPH CZAR OF WESTMINSTER ABBEY MUST HAVE DEMURRED, FOR the great man's grave does not bear these intended words. But Alexander Pope did write a memorable (and technically even heroic) couplet for the tombstone of his most illustrious contemporary. Biblical parodies, perhaps, could not pass muster in Britain's holiest of holies, both sacred and secular,* for Pope's epitome of a life well lived recalled the first overt order of the ultimate boss:

> Nature and Nature's laws lay hid in night:
> God said, let Newton be! and all was light.

Pope surely wins first prize for succinctness and rhyme, but we may cite any number of statements from the wisest of his contemporaries to the best of later scholars, all affirming that something truly special roiled the world of seventeenth-century thinkers, changing the very definitions of knowledge and causality, and achieving a beginning of control over nature (or at least predictability of her ways) that previous centuries had not attained or, for the

---

*Moreover, as a Catholic with a surname that precluded any forgetfulness of such apostasy in Anglican Britain, Pope's popularity at the abbey probably stood about as high as the poet's own stature of four feet six inches!

most part, even sought. Although hard to define, and even denied by some, this transforming period has been awarded the two ultimate verbal accolades by a generally timid profession of academic historians: the definite article for uniqueness, and uppercase designation for importance. Historians generally refer to this watershed of the seventeenth century as the Scientific Revolution.

To cite a key contemporary, a poet rather than a scientist, at least by current disciplinary allocations that would not then have been granted or conceptualized in the same way, John Dryden wrote in 1668:

> Is it not evident, in these last hundred years (when the Study of Philosophy has been the business of all the Virtuosi in Christendome) that almost a new Nature has been revealed to us? That more errors of the School [that is, of the medieval scholastic thinkers and followers of Thomas Aquinas, generally called Schoolmen] have been detected, more useful Experiments in Philosophy have been made, more Noble Secrets in Opticks, Medicine, Anatomy, Astronomy, discovered than in all those credulous and doting Ages from Aristotle to us? So true it is that nothing spreads more fast than Science, when rightly and generally cultivated.

To cite one of the twentieth century's most celebrated philosophers, A. N. Whitehead claimed, in *Science and the Modern World,* that "a brief, and sufficiently accurate description of the intellectual life of the European races during the succeeding two centuries and a quarter up to our own times is that they have been living upon the accumulated capital of ideas provided for them by the genius of the seventeenth century."

A broader range of views could be cited among historians of science, but few would deny that truly extraordinary changes in concepts of natural order—changes that we continue to recognize today as the familiar bases of modern sensibilities—occurred in seventeenth-century Europe, leading to the enterprise that we call "science," with all attendant benefits, travails, and transformation in our collective lives and societies.

In 1939, Alexander Koyré, the dean of twentieth-century students of the Scientific Revolution, described this seventeenth-century transformation as a "veritable 'mutation' of the human intellect . . . one of the most important, if not the most important, since the invention of the Cosmos by Greek thought." The Scientific Revolution, according to the eminent historian

Herbert Butterfield (1957), "outshines everything since the rise of Christianity and reduces the Renaissance and Reformation to the rank of mere episodes, mere internal displacements, within the system of medieval Christendom." And, in 1986, historian of science Richard S. Westfall stated: "The Scientific Revolution was the most important 'event' in Western history. . . . For good and for ill, science stands at the center of every dimension of modern life. It has shaped most of the categories in terms of which we think, and in the process has frequently subverted humanistic concepts that furnished the sinews of our civilization."

In the cartoonish caricature of a "one-line" primer, the Scientific Revolution boasts two philosophical founders of the early seventeenth century—the Englishman Francis Bacon (1561–1626), who touted observational and experimental methods, and the Frenchman René Descartes (1596–1650), who promulgated the mechanical worldview. Galileo (1564–1642) then becomes the first astoundingly successful practitioner, the man who discovered the moons of Jupiter, rearranged the cosmos with a raft of additional telescopic defenses of Copernicus, and famously proclaimed that the "grand book" of nature—that is, the universe—"is written in the language of mathematics, and its characters are triangles, circles, and other geometrical figures." (Galileo's status as martyr to the Roman Inquisition—for he spent the last nine years of his life under the equivalent of "house arrest," following his forced recantation in 1633—also, and justly, enhances his role as a primary hero of rationality.) But the culmination, both in triumphant practice and in fully formulated methodology, resides in a remarkable conjunction of late-seventeenth-century talent, a generation epitomized and honored with the name of its preeminent leader, Isaac Newton (1642–1727), who enjoyed the good fortune of coexistence with so many other brilliant thinkers and doers, most notably Robert Boyle (1627–1691), Edmund Halley (1656–1742), and Robert Hooke (1635–1703).

As with all caricatures based on simplistic historical models of accreting "betterness" (whether by smoothly accumulating improvement or by discontinuous leaps of progress), and on false dichotomies of a bad "before" replaced by a good "after," this description of the Scientific Revolution cannot survive a careful scrutiny of any major aspect of the standard story. To cite just two objections with pedigrees virtually as long as the conventional formulation itself: First, the break between the supposedly benighted Aristotelianism of medieval and Renaissance scholarship, and the experimental and mechanical reforms of the Scientific Revolution, can be recast as far more continuous, with many key insights and discoveries achieved long before the seventeenth century, and abundantly transmitted across the supposed divide. In an early rebuttal that became almost as well known as the basic case for a discontinuous revolution, the French scholar Pierre Duhem, in the opening years of the twentieth century, published three volumes on Leonardo and his precursors. Here Duhem argued that several cornerstones of the Scientific Revolution had been formulated by Aristotelian scholars in fourteenth-century Paris, and had also become sufficiently familiar and accessible that even the formally ill-educated Leonardo, albeit the most brilliant raw intellect of his (or any other) age, sought out and utilized this work, often struggling with Latin texts that he could only read in a halting fashion, as the foundation for his own views of nature. (Duhem developed his thesis under a complex *parti pris* of personal belief, including strong nationalistic and Catholic elements, but his predisposing biases, although markedly different from the *a priori* commitments of historians who built the conventional view, cannot be labeled as stronger or more distorting.)

Second, and in an objection close to the heart of my own persona and chosen profession, the conventional view does seem more than a tad parochial in its nearly exclusive focus on the physical sciences, and upon the kinds of relatively simple problems solvable by controlled experiment and subject to reliable mathematical formulation. What can we say about the sciences of natural history, which underwent equally extensive and strikingly similar changes in the seventeenth century, but largely without the explicit benefit of such experimental and mathematical reconstitution? Did students of living (and geological) nature merely act as camp followers, passively catching the reflected beams of victorious physics and astronomy? Or did the Scientific Revolution encompass bigger, and perhaps more elusive, themes only partially and imperfectly rendered by the admitted triumphs of new discovery and discombobulations of old beliefs so evident in seventeenth-century physics and astronomy? (Because these questions intrigue me, and because my own expertise lies in this area, I shall choose my examples almost entirely from this neglected study of the impact of the Scientific Revolution upon natural history.)

I derived much of the framework, and many of the quotations, for this opening section from the long and excellent treatise of H. Floris Cohen (*The Scientific Revolution: A Historiographical Inquiry,* University of Chicago Press, 1994), a work not so much about the content of the Scientific Revolution as about the construction of the concept by historians. Cohen locates much of the difficulty in defining this episode, or any other major "event" in the history of ideas for that matter, in the complex and elusive nature of change itself. We encounter enough trouble in trying to define and characterize the transformation of clear material entities—the evolution of the human lineage, for example. How shall we treat major changes in our approach to the very nature of knowledge and causality? Cohen writes: "To strike the proper balance between a perception of historical events as relatively continuous or relatively discontinuous has been the historian's task ever since the craft attained maturity in the course of the nineteenth century." The Scientific Revolution becomes so elusive in the enormity of its undeniable impact that Steven Shapin, something of an *enfant terrible* among conventional academicians, opened his iconoclastic, but much respected, study (*The Scientific Revolution,* University of Chicago Press, 1996), with a zinger rich in wisdom within an apparent self-contradiction: "There was no such thing as the Scientific Revolution, and this is a book about it."

We may epitomize the fundamental nature of an episode so fecund in scope and effect, albeit so difficult to characterize, by citing any preferred motto or metaphor in the tradition of "crossing the Rubicon" or "opening Pandora's box." Something tumultuous, permanent, and revolutionary, both for the history of society and the history of ideas, occurred during the course of the seventeenth century. And we may epitomize this extended "event" as the birth pangs and adequate initial development of what we call "modern science," with all its practical consequences for technology, and its intellectual implications for our definition and understanding of "reality" itself. Something happened. Something very big indeed, yet something that we have still not integrated fully and comfortably into the broader fabric of our lives, including the dimensions—humanistic, aesthetic, ethical, and theological—that science cannot resolve, but that science has also (and without contradiction) intimately contacted in every corner of its discourse and being.

Thus, if we wish to understand the continuously troubled relationships between science and these other magisteria of our full being—in this case, and for this book, the interactions between science and the humanities—we would do well to begin at the beginning of modern science, by trying to understand how the seventeenth-century initiators of the Scientific Revolution understood their task, their challenges, their enemies, and their accomplishments. (I discussed the other great pseudo-conflict, the supposed struggle between science and religion, in a previous book, *Rocks of Ages*, Ballantine, 1999.) How, in particular, did these creators of modern science construe the traditional disciplines of humanistic study? How, in even more particular (and to foreshadow a primary theme of this book), did the perception of certain humanistic modes of study as impediments to be swept aside, rather than as allies to be cultivated, set an unfortunate, if understandable (and probably unavoidable), initial context for interaction? Why does this notion of inherent conflict continue to flourish, literally centuries after the growth and success of science destroyed any conceivable rationale for such pugnacity and philistinism? Perhaps a new kid on the block must be scrappy, vigilant, and predisposed to a taxonomy of us against them. But a prosperous and victorious adult should welcome both the moral and the practical obligations of generosity.

My motivation to write this book stems largely from a personal sense of puzzlement. From earliest memories (once I passed through the policeman and fireman stages of universal boyhood, and once I bowed to reality and admitted that I would never occupy center field in Yankee Stadium as a professional address), I wanted to become a scientist "when I grow up"—in particular, once I learned the technical term for folks who study fossils full time, a paleontologist. I could cultivate no immediate family member as mentor or role model, for my closest relatives possessed smarts in abundance, but had not enjoyed access to higher education and professional life. I always loved, for reasons of personal pleasure rather than any "ought" of class or culture, several areas of what traditional taxonomy calls the arts and humanities—from the largely passive delights of reading; to more ambulatory pleasures of a taste for architecture (beats the hell out of birdwatching, if you ask me, for those of a taxonomic bent, as buildings stay put and don't need to be seen at 6:00 A.M. or some other odd time better spent elsewhere); to serious and active participation, still continuing, in choral singing.

I never sensed any conflict among these passions; after all, I seemed reasonably well integrated, at least in my own head and being (my hedgehog side). Indeed, in the naively narcissistic way of childhood, I imagined myself as a perfectly reasonable common denominator of all these activities (my foxy interests). Moreover, lacking direct or familial experience, I didn't even know that science was supposed to conflict with, or even be substantially different from, the arts and humanities.

I did learn the conventional taxonomies later, but they never made any sense to me. I do acknowledge, of course, the historical reasons for conflict—and much of this book, including these opening sections, explores this currently illegitimate basis for suspicion and separation. I also understand that basic pursuits of the sciences and humanities often differ intrinsically and logically, so much so that the techniques of one domain frequently cannot, in principle, answer the questions of the other. In the most obvious example, science tries to ascertain the factual structure of the natural world, whereas criteria for judgment in the arts invoke aesthetic concerns that do not translate into the scientist's language of "true" and "false"—and truth just isn't beauty, however much we may value both, and whatever Mr. Keats found on his Grecian urn. Similarly, and even more broadly—thus providing an even riper bone for false contention when either side misperceives its limits and claims dominion in the other side's magisterium—no factual conclusion of science (a statement about the "is" of nature) can logically determine an ethical truth (a statement about the "ought" of our duties).

Still, all these obvious and well-rehearsed distinctions aside, I have long felt that the similar goals and mental styles overwhelm the legitimate differences in materials for study and modes of validation. The commonalities of creative thinking, and the psychology of mental drive and excitement, seem to transcend the logical differences of subject or approach. (I would not try to distinguish the emotions of exaltation felt in singing a particularly moving passage in Bach's *Passion* settings from the excitement of solving a tough little puzzle in the systematics of *Cerion* [the land snail of my personal research], and saying to myself, "Oh, so that's how it goes!" Late in his life, a celebrated senior colleague stated to me, during a chance encounter on the New York subway of all places, that he continued to love and practice research with all his heart because its pleasures could only be likened to "continual orgasm.")

Moreover, however logically sound and however sanctioned by long historical persistence, our taxonomies of human disciplines arose for largely arbitrary and contingent reasons of past social norms and university practices, thus creating false barriers that impede current understanding. I do not say this to make the obvious point that such boundaries and specializations foster a natural human tendency to jargon and parochialism, but for the much more cogent and useful reason that the conceptual tools needed to solve key problems in one field often migrate beyond our grasp because they become the property of a distant domain, effectively inaccessible to those in need. For example, I feel that I made some breakthroughs in my own field of paleontology only when, remembering the fox's strategy, I explicitly realized that the necessary apparatus for understanding much of life's evolutionary pattern lay in the methodologies established by historians in departments of our humanities faculties, and not in the standard experimental and quantitative procedures so well suited for simple, timeless, and repeatable events in conventional science.

Although the supposed conflict of science and religion has received more press, and has also induced far more actual mayhem over the centuries, the interactions of science with the arts and humanities have been explicitly contested for just as long, and with just as much feeling. In fact—thus my rationale for opening this book at a conventional beginning, that is, at the "official" inception of modern science in the Scientific Revolution of the seventeenth century—this perceived conflict received a canonical formulation right at the starting gate, and has experienced several reincarnations ever since, with different names awarded to the same players, who make the same basic moves (including invention of the same strawman caricatures of the other side), in each "new" episode.

The original version swept through late-seventeenth-century intellectual life as the debate between Ancients and Moderns (with Aristotle and the Renaissance pitted against Bacon and Descartes, as so amusingly depicted by Jonathan Swift, who sided with the Ancients against upstart science, in his deliciously satirical "Battle of the Books"—as discussed in chapter 7). My generation learned the argument primarily through C. P. Snow's widely cited, but rarely read, disquisition on the "two cultures." (Snow, a scientist by training and a novelist and university administrator by later practice, delivered his famous talk on "The Two Cultures and the Scientific Revolution" as the Rede Lecture at Cambridge University in May 1959. He spoke of the growing gap between literary intellectuals and professional scientists, noting for example how "one found Greenwich Village talking precisely the same language as Chelsea, and both having about as much communication with M.I.T. as though the scientists spoke nothing but Tibetan.")

In the years around our millennial transition, scholars resurrected the same debate as the "science wars" between "realists" (including nearly all working scientists), who uphold the objectivity and progressive nature of scientific knowledge, and "relativists" (nearly all housed in faculties of the humanities and social sciences within our universities), who recognize the culturally embedded status of all claims for universal factuality and who regard science as just one system of belief among many alternatives, all worthy of equal weight because the very concept of "scientific truth" can only represent a social construction invented by scientists, whether consciously or not, as a device to justify their "hegemony" (the supposed code word of postmodern cultists) over the study of nature.

This book takes an idiosyncratic, but basically historical, approach to the supposed conflict between science and the humanities by admitting the appropriateness, even the inevitability, of struggle at the birth pangs of modern science, but then arguing that we got stuck, centuries ago, in this superannuated assumption of inherent struggle, when no legitimate rationale—logical, historical, or practical—supports its continuation. Rather, in our increasingly complex and confusing world, we need all the help we can get from each distinct domain of our emotional and intellectual being (the fox's diversity again). *Quilting* a diverse collection of separate patches into a beautiful and integrated coat of many colors, a garment called wisdom (even better than the hedgehog's prickly cover), sure beats *defeating* or *engulfing* as a metaphor for appropriate interaction. My argument proceeds in four statements:

1. Initial conflict between science and the humanities ("the rite and rights of a separating spring" in the title to this first part) inevitably attended the birth of modern science in the Scientific Revolution of the seventeenth century. Big and established boys never cede turf voluntarily, and newcomers must be prepared for a scrappy fight, if only as a ritual of initiation.

2. I shall document, with idiosyncratic examples in the three following chapters, this inevitable and initial estrangement between science and the humanities by asking what the architects of the Scientific Revolution thought they needed to overcome (chapter 2 on a specific example of conflict with humanistic and religious traditions; chapter 3 on issues with the humanities; chapter 4 on the tensions with religious power and orthodoxy). Then, in Part II, I shall show how these founding scientists failed to complete their own mission, in part because they could not do so without some central insights from the humanities.

3. This conflict, initially understandable, became both silly and harmful long ago. Science triumphed in those broad areas rightly belonging to its techniques and expertise. On the other hand, science has no business contending for intellectual turf outside the limits of its stunningly successful methods. Thus the time for peace arrived long ago—and peace would bring such blessings and benefits to both perceived sides, as each has so much to learn from

the successes of the other (as I document by particular examples in chapter 8). Of the famous list of contrasts in the third chapter of Ecclesiastes, we have reached the latter stage in each case: a time to break down, and a time to build up; a time to cast away stones, and a time to gather stones together; a time to rend, and a time to sew; a time of war, and a time of peace.

4. Although such pronouncements may be deemed unfashionable in an age that exalts pluralism and rejects definite solutions, I argue in chapter 9 that we can identify a right and a wrong way to achieve a proper healing of our age-old conflict between science and the humanities. This proper path stresses respect for preciously different insights, inherent to the various crafts, and rejects the language (and practice) of hierarchical worth and subsumption. Consilience, in the definition of the word's inventor, arises from a patchwork of independent affirmations, not by subsumption under an imposed ensign of false union.

2

Scientific "World-Making" and Critical Braking

PERHAPS THERE SHALL NOT ALWAYS BE AN ENGLAND, BUT THIS ISLAND nation does flaunt some impressive examples of stability—an attribute greatly fostered by the arresting fact that no full-scale invasion by foreign forces has overrun the country since 1066. The Oxbridge universities provide several striking examples, including New College, Oxford, named for its inception in 1379 as an upstart among older segments of the university. Similarly, several named chairs have continued for centuries. Stephen Hawking, for example, now serves as Lucasian professor of mathematics, the same chair and title held by Isaac Newton. Cambridge also continues to maintain the prestigious Woodwardian professorship of geology, the first chair in this subject ever established at an English university. Moreover, this title represents more than a merely abstract name, backed by ancient investments still yielding sufficient current interest—for the university's geological museum began its collection with two beautiful cabinets, still intact and proudly on display, built for John Woodward to house his collection of rocks and fossils. The collection also survives therein, essentially complete, and still under the care of the Woodwardian professor.

The life and works of John Woodward (1665–1728), quintessential officer of the Scientific Revolution (not a general like Newton, but not a private either, and one of the most interesting geological thinkers of his day), clearly embody the sensibilities of that interesting time—for the vernacular often surpasses the stellar as a source of insight about the main thrust of a movement, in this case the origins of modern science. Woodward spearheaded an important and popular inquiry within late-seventeenth-century anglophonic science: the attempt to develop comprehensive theories for explaining the complex, and often cataclysmic, history of the earth (a subject previously allocated to the domain of scriptural authority) in mechanical and observational terms favored by the emerging Scientific Revolution—an activity generally known to its detractors, but sometimes worn as a badge of pride by practitioners, as "world-making." Woodward's commitment to the shared precepts of the founders of modern science pervades his work (even though, as we shall see, he ranked among the most pluralistic and least zealous of adherents). First, he accumulated one of the most important and comprehensive collections of rocks and fossils ever assembled in his country, all explicitly done to establish an empirical basis for understanding the structure and history of the earth.

Collecting had been in vogue among scholars and cognoscenti for some time, with several great sixteenth-century museums preceding the development of scientific concerns that would fully emerge only in Woodward's time. For example, Peter the Great, Woodward's regal contemporary, built one of the finest and most extensive collections in Western history (largely by direct purchase rather than personal gathering, given his circumstances of maximal resources and minimal time). Much of this material still survives, on display in the Kunstkamera of St. Petersburg, the building expressly constructed by the czar to house his spoils. But these earlier collections, as reflected in the general name for their display sites—*Wunderkämmern,* or rooms of wonders—express the different aims and sensibilities of Renaissance and Baroque times: to evoke visceral awe at nature's diversity; to flaunt the rare and the bizarre (and to outshine other collectors) by owning the strangest and the superlative (the oddest or most deformed, the largest, the most beautiful); to mix both objects of nature and products of human design, thus to sample everything of interest, as promiscuously as possible. But under the transformed agenda of the Scientific Revolution, collectors became more interested in fathoming nature's order than in eliciting human awe; in developing museums that would reveal, in actual objects, the history and system of nature's lawlike ways.

Second, Woodward's literary efforts in "world-making," his writings on the nature of fossils and the history of the earth, stress the themes of Newton's generation, primarily the power of observational and experimental methods to gain new and reliable knowledge inaccessible to former styles of scholarship. Woodward's most famous book, *An Essay Toward a Natural History of Earth,* published in 1695, illustrates the emerging consensus, even though the peculiar (and quite incorrect) content of Woodward's central theory might lead us to dismiss his work, with superficial scorn, as a paragon of prescientific irrationalism.

We can learn a great deal from the covert content of actual, and apparently insignificant, illustrations. Note, for example, the broad aims of world-making, as expressed in the ample heading of Woodward's title page (figure 1). But the far fewer words of the facing left-hand page (figure 2) speak volumes about the hold of the old, and the embattled pugnacity of the new. I argue throughout this part that these earliest modern scientists, in the throes of their birth pangs, worried about the power and misleading ways of two entrenched and prestigious institutions: official theology (but not religion per se, as all these men were personally devout and seriously Christian), and the hidebound traditions of humanistic scholarship.

This left-hand page represents a sly dig at the first impediment. Books published under Catholic auspices had to pass the scrutiny of official censors, and receive an imprimatur—literally meaning "let it be printed." This imprimatur had to be attested, signed, and published, usually on the left-hand page just before the start of the book's actual content. The presses of Anglican Britain operated under no such restriction, so Woodward displayed the greater liberty of his culture with an "official" imprimatur signed not by a theological guardian of doctrine and morality, but by John Hoskyns, described as V.P.R.S., or "vice president of the Royal Society," Britain's primary institution for the promotion of science.

As another emblem of changes not yet made, and tension between old and new, why does the date of the imprimatur appear dually and ambiguously as "Jan. 3, 1694/5"? Couldn't the august Royal Society read a calendar and know the correct year? The Gregorian reform had properly reset calendars to acknowledge the year's true length as 365 and not-quite-a-quarter-of-an-extra-day (requiring the suppression of occasional leap years, rather than adding the extra

An ESSAY toward a

Natural History

OF THE

EARTH:

AND

Terreſtrial Bodies,

Eſpecially

MINERALS:

As alſo of the

Sea, Rivers, *and* Springs.

With an Account of the

UNIVERSAL DELUGE:

And of the *Effects* that it had upon the

EARTH.

By *John Woodward*, M. D. Profeſſor of Phyſick in *Greſham-College*, and Fellow of the *Royal Society*.

LONDON: Printed for *Ric. Wilkin* at the *Kings-Head* in St. *Paul's* Church-yard, 1695.

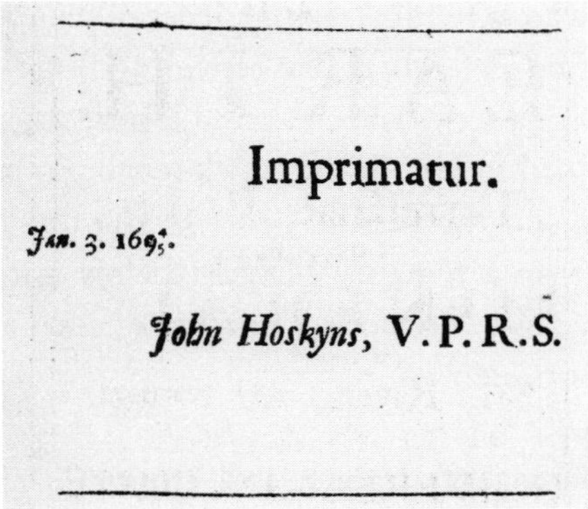

Imprimatur.

Jan. 3. 169⅘.

John Hoskyns, V. P. R. S.

Figure 2.

day every four years, as in the Julian calendar, used ever since Caesar named the system for himself). But the new calendar had been proclaimed by Pope Gregory in 1582 and therefore smelled like a papist plot to the Anglican Brits (who did not succumb to astronomical reality and adopt the Gregorian system until the mid-eighteenth century). Now, the Julian calendar, in addition to falling so out of whack with the solstices and equinoxes as its small errors accumulated over the centuries, also began the year on March 1. Thus the January 3 of Woodward's imprimatur fell within 1694 on the Julian calendar still used in England, but within 1695 on the more accurate Gregorian calendar (which had also switched the admittedly arbitrary beginning of the year to January 1).

What, then, was a budding English scientist to do—be patriotic and parochial, or scientific and universal? (The same dilemma continues to face American scientists in the only Western nation that has not adopted the scientifically far more convenient metric system. My colleagues and I would not dream of using anything but metric units in our technical publications, but what should we do in our popular articles? Moreover, ambiguity can be expensive, as illustrated by the recent failure of an important mission to Mars, triggered by a vernacular figure that engineers had read as metric.) In any case, Woodward cited both alternatives, letting his readers pay their money for his book and then make their choice.

In his distinctive brand of world-making, Woodward attributed the form and sedimentary constitution of our present earth almost entirely to Noah's flood, which, according to his theory, had dissolved all nonorganic material of

Figure 1.

the original earth into a sludge of oceanic waters (while leaving pieces of plants and animals more intact for later embedding as fossils). The retreating flood-waters then deposited this universal slurry as a series of horizontal layers, thus forming the earth's strata with their contained fossils. In the quick settling out of this slurry, the heaviest fossils fell first, and the vertical order of the fossil record therefore records the density of organic remains, heaviest below and lightest on top! Woodward confessed the admittedly peculiar nature of such a "wild" claim in his book's preface and first presentation:*

> It will perhaps at first sight seem very strange, and almost shock an ordinary reader to find me asserting, as I do, that the whole terrestrial globe was taken all to pieces and dissolved at the Deluge, the particles of stone, marble, and all other solid [rocks] dissevered, taken up into the water, and there sustained together with sea-shells and other animal and vegetable bodies; and that the present earth consists, and was formed out of that promiscuous mass. . . . That the said terrestrial matter is disposed into strata or layers, placed one upon another in like manner as any earthy sed-iment, settling down from a fluid in great quantity will naturally be; that these marine bodies are now found lodged in those strata according to the order of their gravity, those which are heaviest lying deepest.

---

*Seventeenth-century English used a very flexible and often inconsistent system of spelling. These texts also capitalized many words that would not receive such treatment in modern English, and used far more punctuation (particularly placing commas and semicolons where we would use no punctuation marks). Any scholar faces a dilemma in trying to devise a "best way" for presenting such quotations in general works for modern readers. I have opted for the decision most commonly adopted by my colleagues. I see no sense in slavishly copying the original spellings (which are inconsistent in any case). Thus I use modern spellings, punctua-tions, and capitalizations. But I strictly follow the actual wording, even when archaic (which I generally find charming in any case). If an archaic wording might confuse modern readers, I retain it anyway and add an immediate explanation in brackets. This procedure, I believe, retains the full flavor and entirely accurate wording of the original, while only modernizing the fluid conventions of typography. (In the few cases of citation from a modern secondary source, rather than from the original document—as in the lines from Dryden on page 12, taken from H. Floris Cohen's 1994 book, *The Scientific Revolution: A Historiographical Inquiry*—I have reprinted the quotation exactly as I found it in the secondary text—hence the archaic retentions in the Dryden quote, because Cohen so presents it.)

At first hearing, any modern geologist might be forgiven for assuming that Woodward must rank among the antiscientific theological dogmatists, committed to validating the literal truth of scripture, inventing fanciful explanations from an armchair, and caring not a whit about the actual character of fossils or the sedimentary record. I must also confess that this first impression receives strong enhancement from the remarkable, if entirely incidental, fact that one of the leading "theories" purveyed by modern American fundamentalists who call themselves "creation scientists"—one of the great oxymorons of our time—promotes the exact same explanation for the fossil record: as the product of a single event that occurred during the few thousand years of allotted biblical time and left a paleontological record ordered not by the millions of years needed to evolve new species from their predecessors, but only in the few years required to deposit all fossils in order of their density—a patent empirical absurdity since some of the lightest and most delicate fossils occur in the oldest strata, while many massive forms (mammoth teeth and bones for starters) can only be found at the top of the stratigraphic pile. (See, for example, the "bible" of modern creationism— *The Genesis Flood,* by J. C. Whitcomb and H. Morris.)

But Woodward strongly rejects this admittedly plausible claim that he operated as an armchair theologian, and situates himself squarely within the developing beliefs and procedures of the Scientific Revolution, insisting that, however strange or improbable his theory may appear, he had reached his conclusions by inference from copious observations of fossils and strata, and that the theory would survive or expire on the strength and validity of these and subsequent empirical studies. Indeed, the opening words of his treatise proclaim:

> From a long train of experience, the world is at length convinced that observations are the only sure grounds whereon to build a lasting and substantial philosophy. All parties are so far agreed upon this matter that it seems to be now the common sense of mankind. For which reason, I shall in the work before me, give myself to be guided wholly by matter of fact . . . and not to offer anything but what hath due warrant from observations; and those both carefully made and faithfully related.

(Needless to say, I am not arguing that Woodward strictly followed his own stated ideals, because no one can—and theoretical preferences always intrude, even when our most honorable intentions lead us to believe most fervently that we only follow the objective dictates of pure observation.)

Woodward even maintains that the oddity of his theory, forced upon him as the only way to explain his observations, weighed heavily upon his sense of rationality:

> In truth the thing, at first, appeared so wonderful and surprising to me, that I must confess I was for some time at a stand; nor could I bring over my reason to assent, until, by a deliberate and careful examination of these marine bodies, I was abundantly convinced that they could not have come into those circumstances by any other means than such a dissolution of the earth, and confusion of things. And were it not that the observations, made in so many, and those so distant, places, and repeated so often with the most scrupulous and diffident circumspection, did so establish and ascertain the thing, as not to leave any room for contest or doubt, I could scarcely even have credited it.

How, then, did Woodward and other prominent "world-makers" of his time characterize the major forces and impediments aligned against the new scientific approach to establishing firm, and ever increasing, knowledge about the ways and means of nature? Two themes and worries circulate throughout the writings of Woodward and other world-makers. These two themes also set the foundation for this book, and for my key claim that the oppositional pugnacity, so understandable and necessary in the initial defense of a fledgling movement against genuine adversaries and powerful inertia, became unseemly and inappropriate ages ago.

First, Woodward railed against the humanistic tradition, which had so singularly failed to comprehend the true nature of fossils as organic remains because these prescientific scholars had used false criteria and classifications based on human needs and perceptions, rather than basing their schemes of order upon the natural and mechanical status of the objects themselves. Woodward wrote, in his great posthumously published work of 1728, *Fossils of All Kinds, Digested into a Method, Suitable to Their Mutual Relation and Affinity:*

> In their [the humanists'] methodizing and ranging of the native fossils, 'tis no wonder that they fail, and that all things are in disor-

der, and out of course with them, when they so frequently make choice of characters, to rank them by, that are wholly accidental, and unphilosophical; as having no foundation in nature, or the constitution of the bodies themselves. Thus some rank them under the heads of common, and rare, or mean and precious; of less, and of greater use. Then they reduce them to subordinate classes, according to their particular uses, in medicine, surgery, painting, smithery, and the like; which would be proper in an history of art, or mechanics; but serves only to mislead them and their readers in the history of nature.

Second, the doubts, imprecations, and even frontal assaults of parts of the theological establishment, and of some conservative religious thinkers, posed a threat to the freedom of emerging scientific practice to seek mechanical explanations based on natural laws, and to explore an expanded concept of the earth's age. I cannot emphasize too strongly that the old model of all-out warfare between science and religion—the "standard" view of my secular education, and founded upon two wildly successful books of the mid- to late nineteenth century (Draper, 1874, and White, 1896; see my book *Rocks of Ages,* previously cited [page 16], for more on the fallacies of this construct)—simply does not fit this issue, and represents an absurdly false and caricatured dichotomy that can only disrespect both supposed sides of this nonexistent conflict. "Religion," as a coherent entity, never opposed "science" in any general or comprehensive way.

Some dogmatists and traditionalists did greatly fear the influence of science and did hold strongly to biblical literalism (particularly for the earth's young age, a short episode of creation, and the reality of a universal flood) and to the impossibility of explaining planetary history without frequent and direct miraculous intervention by God. But many other equally devout and equally professional theologians welcomed the new knowledge as embodying a more exalted concept of God's awesome power and sense of order, and of his wisdom in establishing a universe operating under constant laws of his own proclamation—regularities that would require no divine tampering thereafter. Moreover, effectively all major scientists of the seventeenth century held genuine and firm beliefs in a deity of fairly conventional form, and in the sacredness and infallibility (if not the literality) of the Bible.

Indeed, the movers and shakers of the Scientific Revolution espoused a range of opinions about even so central a question as whether miracles—defined as direct supernatural interventions of God—must be invoked to explain the full range of phenomena required to chronicle the earth's history. (In the practical terms of a working scientist, miracles must be treated as temporary suspensions for the otherwise invariant regime of natural laws that regulate the mechanisms of the universe, and make scientific explanation possible.) Newton himself took a "generous" view, and happily granted God an occasional option for miraculous intervention, even though the success of science did imply God's strong preference for invariant natural law, and his sparing use of miraculous effects. Newton wrote: "Where natural causes are at hand God uses them as instruments in his works, but I do not think them alone sufficient."

But other key figures of Newton's generation hoped to ban miracles entirely, arguing that God would only be demeaning his own grandeur if he ever needed to fiddle, even momentarily, with his own laws to nudge history back into the right track. And just to prove that degree of theological commitment did not necessarily correlate with warm feelings toward occasional miracles, the leading Anglican theologian and naturalist, the Reverend Thomas Burnet, became the staunchest defender of full sufficiency for ordinary physical laws, and the strongest opponent of miracles, even for explaining such spectacular events as Noah's flood. Burnet's *Sacred Theory of the Earth*, first published in 1680, became the most influential (and controversial) document in the growing scientific literature on "world-making." (Newton's brief for occasional miracles, quoted above, comes from a letter he wrote to Burnet, criticizing his close friend and scientific colleague for insisting that God must restrict himself to ordinary physical processes in the extraordinary task of world-making.)

Woodward followed Newton's lead on this issue, and frankly admitted that he could imagine no mechanism for dissolving the original earth into the slurry of Noah's flood except a miraculous suspension of gravity, causing particles that normally cohere to fall apart and dissolve. But—and here we grasp the crucial difference between scientists who permitted occasional miracles and traditionalists who embraced divine intervention as the preferred and primary motor of nature's substantial events—Woodward only called upon miraculous agency when natural explanation had, in his view, clearly and irretrievably failed, and when observation plainly implied the existence of phenomena that could not be explained in any other way. In short, observational necessity (rather than theoretical or theological preference) established the only acceptable ground for invoking divine intervention—an explanation of last resort, to be acknowledged only when invariant natural law simply could not, in his understanding, generate an empirically affirmed set of results.

In reading Woodward's major work of 1695, one senses that he regarded his appeal to miraculous agency for dissolving the earth into the floodwaters as troublesome to the spirit of developing scientific methods, and more than a bit embarrassing. For, in describing this central argument of his entire theoretical apparatus, Woodward ventures only two short statements, one a quick admission, and the other a near apology. He situates the first admission within an explicit critique of Burnet's views on fully physical explanation:

> That the deluge did not happen from an accidental concourse of natural causes . . . That very many things were then certainly done, which never possibly could have been done without the assistance of a supernatural power. That the said power acted in this matter with design, and with higher wisdom. And that, as the system of nature was then, and is still, supported and established, a deluge neither could then, nor can now, happen naturally.

(I may be imposing modern sensibilities upon Woodward's different intentions and motives, but these quick "admissions," presented as a set of minimal statements in partial sentences, do seem to project Woodward's discomfort at his perceived need to invoke a miraculous moment of antigravity as the only means he could devise for explaining a set of primary empirical observations.)

Woodward's second statement, from the introduction to his book, comes very close to outright apology for calling upon miracles against the preferences of science for natural explanation:

> A few advances there are in the following pages, tending to assert the superintendence and agency of Providence in the natural world. . . . But I may very safely say, that . . . I have not entered farther into it than merely I was led by the necessity of my subject: nor could I have done less than I have, without the most apparent injury and injustice to truth.

The genuine tension between developing scientific and traditional scriptural explanations of world-making bursts forth in the strongly polemical writings of John Keill (1671–1721), Savilian professor of astronomy at Oxford, and the most theologically conservative scholar in Newton's immediate circle. In his article for the *Dictionary of Scientific Biography*, David Kubrin described Keill as "one of the few around Newton with High Church patronage." Keill's brief for miracles differed radically from the careful and minimal claims of Newton or Woodward. Keill reviled the Cartesians as prophets of atheism (albeit unintentionally, he grudgingly admitted) in their claims that God restricted himself to natural laws established at the creation—for a deity so hampered in direct expression of his singular powers might just as well not exist at all. In contrast with Woodward's caution, Keill defended a muscular preference for frequent miracles as the preferred and proper way of God, the primary agent of major episodes in nature's history, and also the most potent weapon against the dangerous atheistic tendencies of his age. In Keill's view, Kubrin continues, "natural theology should be subordinated to the Scripture, while natural philosophy should acknowledge the important role played not only by Providence but also by outright miracles."

In his major geological treatise of 1698, for example, Keill presents an explicit defense of miracles as the primary shapers of major events in the earth's history, thereby placing this central aspect of the natural world permanently outside the province and possibility of scientific understanding:

> The scriptures give us an account of several miracles wrought by the hand of Omnipotence upon several occasions, which did not so necessarily require them. Why ought we then to deny this universal destruction of the earth [the deluge of Noah] to be miraculous? Miracles are the great and wonderful works of God, by which he showeth his Dominion and Power, and that his Kingdom reacheth over all, even Nature herself, and that he does not confine himself to the ordinary methods of acting, but can alter them according to his pleasure. Were not they [miracles] given us to convince us of the sacred truths contained in holy scripture? . . . Certainly we are not to detract from the value of [miracles] by pretending to deduce them from natural and mechanical causes, when they are by no ways explicable by them. It is therefore both the easiest and safest way to refer the wonderful destruction of the old world to the Omnipotent hand of God, who can do whatsoever he pleases.

If the efficacy of scientific explanation could be denied to such a central subject in natural history by a colleague serving as professor of astronomy at his nation's greatest (if rather hidebound) university, then what greater imprecations might flow from professional humanists and theologians with genuine and pervasive animus against a fledgling enterprise, science itself, that (to mix both metaphors and taxonomic categories) might yet be nipped in the bud of its initial promise? To cite just one prominent example of why a budding scientist might reasonably entertain such a fear, consider some words of warning from Edward Stillingfleet, a famous scholar and conservative theologian who, in 1662, published his defense of scriptural primacy in a popular work with a forthright title: *Origines Sacrae, or a Rational Account of the Grounds of Christian Faith, as to the Truth and Divine Authority of the Scriptures, and the Matters Therein Contained.* (I quote from my copy of the 1666 edition, printed by Henry Mortlock "at the sign of the Phoenix in St. Paul's Church-yard near the little north door," just a few months before the great fire of London burned that region of town to rubble in a truly nonmiraculous catastrophe.)

Stillingfleet vented his harshest criticism against the standard theological construct of Newtonian scientists: the "clockwinder" God who got everything right at the creation and had never again imposed his direct hand upon the works of nature. Such an abstract and distant deity, Stillingfleet insists, cannot serve as a satisfying source for our veneration, our moral rectitude, or our hope of eternal reward. The rise of the mechanical philosophy, and the Scientific Revolution in general, therefore truly threatens our psychological safety and public order:

> What expressions of gratitude can be left to God for his goodness, if he interpose not in the affairs of the world? . . . For if the world did of necessity exist, then God is no free agent; and if so, then all instituted Religion is to no purpose; nor can there be any expectation of reward, or fear of punishment from him who hath nothing else to do in the world but to set the great wheel of the heavens going.

In short, the leaders of the Scientific Revolution did encounter genuine intellectual opponents of no mean force, powerful critics who held all the advantages of incumbency and the weight of tradition. One can scarcely blame science for a little pugnacity in its infancy.

3

So Noble an Hecatombe:
The Weight of Humanism

THE PECULIAR NOTION THAT SCIENCE UTILIZES PURE AND UNBIASED observation as the only and ultimate method for discovering nature's truth, operates as the foundational (and, I would argue, rather pernicious) myth of my profession. Scientists could not so approach the world even if we justly so desired—for, as the distinguished philosopher of science N. R. Hanson once remarked, "the cloven hoofprint of theory" necessarily intrudes upon any scheme of observation. So must it be, and so should it be—for how could we ever discern a pattern, or see anything coherent, amid an infinitude of potential perceptions, unless we employed some theoretical expectation to guide our penetration of this plethora. Bias cannot be equated with the existence of a preference; rather, bias should be defined as our unwillingness to abandon these preferences (or at least to challenge them further and rigorously) when nature seems to say "no" to our explicit searches and tests. Indeed, most scientists distinguish their work by imposing a conscious and opposite bias upon their practice—that is, by applying *greater* skepticism and *more rigorous* (and frequent) testing to observations that support their preferences—precisely because they know how enticing, and thus how impervious to refutation, such preferences can become.

Thoughtful scientists have always recognized both the philosophical necessity and the practical advantages of observations made to test theoretical preferences, rather than promiscuously recorded as random items of a mindless list. In one of my favorite "great quotations," Charles Darwin wrote to a close colleague about the myth of "objective" recording: "How odd it is that anyone should not see that all observation must be for or against some view if it is to be of any service."* In this context, if we wish to grasp the major intentions of the leaders of the Scientific Revolution, we ought to inquire about their opposition, at least as they perceived and depicted the field of battle. What folks, and what ideas, did they brand as their primary impediments—for all observation must be "for or against" some view. As stated above, I wish to bypass the canonical heros from Galileo to Newton and focus instead upon the leading taxonomists and naturalists of Newton's generation—both because my expertise lies in this area and because we have more to learn from the neglected than from the overly eulogized.

I shall begin with a quotation that makes perfect sense in a late-seventeenth-century context, but would strike any modern scientist as decidedly peculiar. I encountered this statement in a famous work by Britain's finest naturalist of Newton's time. John Ray (1627–1705)—the great taxonomist of plants, birds, fishes, and fossils, the man who even published a compendium of proverbs to show that human aphorisms could be classified by the same principles used to arrange organisms—wondered in his *Synopsis Methodica Animalium Quadrupedium et Serpenti Generis* (A Methodological Synopsis of the Kinds of Four-Footed Animals, and of Serpents), published in 1693, whether his generation could still discover anything new or interesting about animals: *Quid attinet de Animalibus plura scribere? Imo post Aldrovandum et Gesnerum quid scribendum restat?* ("What more can be undertaken in writing about animals? What, indeed, remains to be written after [the work of] Aldrovandi and Gesner?")

This statement sounds so peculiar to modern sensibilities because such a dilemma would never occur to a scientist in the twenty-first century. How could anyone even imagine that everything worth doing had already been

---

*I confess that some of my affinity for this statement rests upon its irrelevant but peculiar property of including six words in a row with only two letters each.

done, or worth knowing already known? Science, after all, can almost be defined by its skepticism and constant probing, by the conviction, always fulfilled so far, that new pathways to the fundamentally unknown still lie before us, however invisible to our current perception and instrumentation.

To grasp Ray's statement, we need to recognize that the leading intellectual movement of the previous century (with Aldrovandi and Gesner as chief exponents in natural history), and the focus of Ray's opposition in 1693, would have depicted the fate of any future zoological study in precisely these terms. Ray therefore tried to engage his opposition, to secure their understanding (and perhaps even their eventual agreement), by posing his alternative in their language.

We often get befuddled when we try to comprehend the central belief of the system that the Scientific Revolution hoped to replace, because this precept strikes us as so strange and archaic, whereas the movement itself still commands our maximal respect for the fame and valor of its heroes and the honor attached to its name. We continue to revere the Renaissance—literally meaning "the rebirth"—because we so admire the works of Leonardo and Michelangelo, and because we so respectfully refer to rare folks of multiple academic talents as "Renaissance" men and women. Thus the very mention of the Renaissance tends to evoke an image of forward-looking modernity in our minds.

But our modern concept of science could never have flourished, or even been conceived at all, under the aims and sensibilities of the Renaissance. This movement did wish to augment the storehouse of human knowledge—but not by discovering new truths of nature through observation and experiment. As the name implies, Renaissance scholars sought a rebirth, not an accumulation or a revolution. They believed that everything worth knowing had been ascertained by the great intellects of the classical world (the glory of Greece, the grandeur of Rome, and all that), but then either not transcribed or, more likely, lost to the West during a thousand intervening dark years, as libraries burned and decayed, and new dogmas clouded the ancient spirit of liberal learning. Thus, for the Renaissance, the *re*covery of ancient wisdom, not the *dis*covery of novel data, became the primary task of scholarship. Returning, then, to Ray's challenge: the belief that rigorous observation of nature might offer only limited utility, or even prove counterproductive—because great scholars (inspired by ancient sources) had already been there and done that—seemed entirely reasonable to champions of the Renaissance, especially given their lukewarm feelings about this tough new kid on the block, who paid superficial homage but who (one suspected) really wished to assume the reins himself, and to discredit Greece and Rome.

As the culmination of their art and the intention of their movement, Ulisse Aldrovandi (1522–1605) of Bologna and Konrad Gesner (1516–1565) of Zurich, the premier natural historians of the Renaissance, published massive and lavishly illustrated compendia on the inhabitants of all three realms of nature—animal, vegetable, and mineral. But the form and purpose of these amazing volumes would strike any modern scientist as surpassingly strange, albeit wonderfully weird. Aldrovandi and Gesner displayed no rooted antipathy to novel information of their own discovery, or to observing animals with their own eyes and recording the results, but such activities represented a diversion from their primary purpose: to transmit everything ever known, stated, or merely believed about the objects under their scrutiny. For example, Gesner's seminal work, the *Historia animalium* (volume 1, on mammals) of 1551, includes an alphabetical series of chapters from *De alce* (on the elk) to *De vulpe* (on the fox), with each entry structured as a compendium of everything ever recorded about the species at hand, with pride of place, and maximal length of treatment, granted to the claims of classical authors, particularly Aristotle for Greek learning, and Pliny (who died in his boots, and with unbeatable panache, in the eruption of Mount Vesuvius in A.D. 79) for the Roman follow-up.

Aldrovandi and Gesner also showed no aversion to true information versus the claims of legends and fables—and they did try to make the distinction in their texts. But facts gained no preferred status by their documentable veracity. After all, these Renaissance scholars viewed completeness of previous human claims and beliefs, not the separation of fact from fiction, as their ultimate goal—for, by their lights, the Ancients had developed the correct framework for everything worth knowing, and our modern efforts must therefore be directed to recovering this knowledge and coordinating Ancient convictions with later claims and beliefs, all the better to compile a full account of human experience with each of God's creatures. (Thus Ray's rhetorical question could not have been more apropos for his time: had Aldrovandi and Gesner left anything unsaid? Or, using this book's metaphor, had the fox's full range already been specified?)

To understand Gesner's Renaissance motives and intentions, one must read his chapters *De monocerote* and *De satyro* ("on the unicorn," and "on the satyr"), both included among genuine mammals not because Gesner credited their actual existence, but because he conceived his work as a full compendium of human attitudes and beliefs about four-footed beasts—and the anatomical attributes, or even the reality of their being, did not figure prominently among his criteria for inclusion. For example, Gesner's thirty-six-page article *De sue* (on pigs), followed by a further fifteen pages *De apro* (on boars), begins with a lengthy discussion of classical knowledge (adorned with copious quotations in Latin and Greek), followed by sections on etymology, gastronomy, pigs as emblems and metaphors in literature, and a list of all recorded proverbs about pigs. Those who would laugh at Gesner, or smile benignly upon his misguided intentions, because he included such *chozzerei* in a supposedly scientific treatise, should revise the assumption behind their smug dismissal. Gesner purposefully assembled all this pig paraphernalia—the more the better, and the greater the admiration that would therefore flow to his thoroughness as a scholar.

If we look to volume *H* of the first edition of the *Oxford English Dictionary*, published in 1901, well before later academic debates of the twentieth century would alter the meanings into code words for partisans of struggle, we find the classical definition of humanism that struck the developing Scientific Revolution as so problematical: "Devotion to those studies which promote human culture; literary culture; *esp.* The system of the Humanists, the study of the Roman and Greek classics which came into vogue at the Renaissance." The humanities, as scholarly subjects, then become "learning or literature concerned with human culture: a term including the various branches of polite scholarship, as grammar, rhetoric, poetry, and *esp.* the study of the ancient Latin and Greek classics." Finally, the *OED* defines a humanist as "one devoted to or versed in the literary studies called 'the humanities'; a classical scholar; *esp.* a Latinist, a professor or teacher of Latin."

Now, nothing in these dry definitions would have offended the nascent scientists of Newton's world—so long as these devotees of Latin, and true believers in the full sufficiency of ancient knowledge, stuck to their literary lasts and did not insist upon applying the same principles, and the same moral ordering of worth among scholarly pursuits, to the study of life and the earth. But, as the leading examples of Gesner and Aldrovandi show so well, the Renaissance humanists did assume that their style of learning applied with equal force and exclusivity throughout the domain of subjects that we now allot to scientific explanation—thus sowing seeds of conflict when the new observational methodologies, with their touch of philistinism toward antique sources of information, challenged the old compendia, long regarded as complete and unsurpassable by their Renaissance champions.

This older explanatory world of Renaissance recovery had to clash with the scientific sensibilities inspired by Bacon and Descartes. For the two schools could hardly have differed more in their definitions of both goals and methods. (1) For goal, the Renaissance desire to refine knowledge by recovering the old versus the scientists' intention to extend knowledge by observing the heretofore unseen, whether by finding unknown objects in previously unexplored lands, or by inventing instruments that could focus and measure the previously unseeable (with the microscope and telescope as primary examples). (2) For method, both Renaissance scholars and budding scientists favored the development of great collections, housed in museums. But the two schools conceived museums as fundamentally different kinds of places, imbued with different purposes—the Renaissance as a complete repository of objects, both natural and manufactured, and dedicated to the compendiast's dream of recording all forms of interaction between human and natural productions. The early scientific museums, on the other hand, rejected such promiscuous ingathering and sought instead to include certain kinds of objects (and to reject others), arranged in an order that would shed light upon the causes and purposes of their natural origins and utilities (in other words, more of a hedgehog's restriction to the display of nature's objective, sensible, coherent, and factual order, independent of human preferences and interactions).

As secretary of the Royal Society, Britain's premier scientific institution (incorporated under a royal charter in 1662 by the restored King Charles II), Nehemiah Grew (1641–1712), best known for his early microscopical studies of plant anatomy, prepared a catalog of the Society's growing collections, published in 1681 under the commodious title *Musaeum Regalis Societatis, or a Catalogue and Description of the Natural and Artificial Rarities Belonging to the Royal Society . . . Whereunto Is Subjoyned a Comparative Anatomy of Stomachs and Guts.* (Waste not, want not—so Grew took this opportunity to include the plates and descriptions of his prototypical research project in the new empirical spirit of the Scientific Revolution: careful dissections, and attempts to understand the resulting differences in functional terms, of digestive systems in a wide variety of vertebrates—see figure 3.)

The pugnacious preface of Grew's catalog clearly designates the "against" within his movement's particular version of "all observation for or against some view." Referring to the Royal Society's collection as "that so noble an

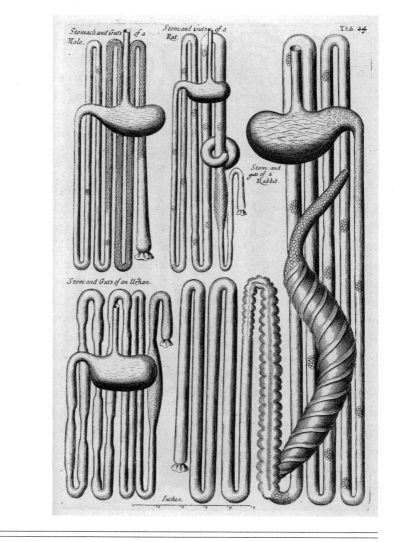

Figure 3.

hecatombe," Grew begins by assuring readers that he came to praise this sacrifice of nature's living bounty by making truthful observations about the skeletons and species so represented, not to bury the plethora of specimens in an old-style Renaissance compendium of promiscuous gobbledygook, arranged only by some scheme of purely human convenience, and not by any natural principle of objective order. He begins by contrasting his taxonomy with the devices of the traditional bugbears, the Renaissance compendiasts Aldrovandi and Gesner:

> I like not the reason which Aldrovandus gives for his beginning the History of Quadrupeds with the Horse; *Quod praecipuam nobis utilitatem praebeat* [because its particular utility to us stands out]. [Organisms are] better placed according to the degrees of their approximation to human shape and one to another: and so other things, according to their nature. Much less should I choose, with Gesner, to go by the alphabet. The very scale of the creatures is a matter of high speculation.

As Grew enumerates the innovations of his scientific colleagues and criticizes the practices of the old compendiasts, the primary theme for this section of my book emerges: the humanistic tradition of the Renaissance had stymied the development of natural history by making the literary claims of ancient writers more important than direct observation of the actual species supposedly under analysis, and by privileging the fables and legends attached to these creatures, largely because such portrayals could be traced to an Aesop or an Aristotle, over new sources of physiological or anatomical information that could explain the biological origin and purpose of observed forms and behaviors.

Grew begins by explicitly damning Pliny with faint praise or, more accurately, by damning those naturalists so committed to a humanist belief in the superiority and sufficiency of Ancient texts that they spend all their time arguing fruitlessly over Pliny's short and cryptic remarks, when they could be observing the relevant organisms by themselves, and making proper decisions with their own eyes and minds. Grew and his scientific colleagues would loudly proclaim that the Plinian emperor wore no clothing, and that the little boy of nascent science could do far better by favoring simple observation over slavish obeisance:

> The curiosity and diligence of Pliny is highly to be commended. Yet he is so brief, that his works are rather a nomenclature than a history: which perhaps might be more intelligible to the age he lived in, than the succeeding ones. But had he, and others, been more particular in the matters they treat of, their commentators had [Grew here uses the old subjunctive mood, equivalent to our modern "would have"] engaged their own and their readers' time much better than in so many fruitless and endless disquisitions and contests.

Grew then states his own support for an entirely new approach, contrary to the aims of Renaissance humanism, and within the spirit of developing science. He suggests two guiding rules, each opposite to the thrust and practice of the humanist tradition: (1) make distinctions between true and false claims rather than writing down everything ever said in a complete accounting of all opinions previously expressed; and (2) base these distinctions on direct observation rather than respect for classical pronouncements:

> It were [another subjunctive, equivalent to our modern "would be"] certainly a thing both in itself desirable, and of much consequence, to have such an inventory of nature wherein, on the one hand, nothing should be wanting, but nothing repeated or confounded on the other. For which, there is no way without a clear and full description of things.

Grew then adds two further criticisms of the humanist tradition, as expressed by the Renaissance compendiasts Gesner and Aldrovandi, before restating his simple recipe in solution. First, unlike them, he will not clutter his text with the paraphernalia of humanistic footnotes that bear no relationship to the *natural* history of organisms under observation:

> After the descriptions, instead of meddling with mystic, mythologic, or hieroglyphic [meaning occult in general, not specifically Egyptian] matters, or relating stories of men who were great riders, or women who were bold and feared not horses, as some others have done, I thought it much more proper to remark some of the uses and reasons of things.

Grew adds that he will also not waste space showing readers his humanistic erudition and knowledge of classical sources, when such quotations only express obvious points that any person could observe for himself, and therefore add nothing to our knowledge of the animal under description. He specifically criticizes Aldrovandi for stating that sheep belong to the group of cloven-hoofed quadrupeds, not only because one can easily see so for oneself, but because *Aristoteles etiam scripto publicavit* (Aristotle also said so in his writing). Grew adds: "I have made the quotations not to prove things well known to be true, as one [he then explicitly names Aldrovandi in a footnote] who very formally quotes Aristotle to prove a sheep to be among the Bisulca [cloven-hoofed beasts]."

Finally, Grew defends his own procedure. He will work by observation, favoring his own studies of living organisms. And he will also provide basic data of weights and measures:

> In the descriptions given, I have observed, with the figures of things, also the colors, so far as I could. . . . And I have added their just measures, much neglected by writers of natural history.

To indicate that Grew expressed a consensus among the early scientists of Newton's circle, and did not state his criticisms as an angry and idiosyncratic Miniver Cheevy, pouring his vitriol upon the world, we should also return to John Ray (1627–1705), the preeminent British naturalist of this founding generation. (Linnaeus himself, in the first edition of his *Systema Naturae* [1735], the work that initiated modern taxonomy, singled out *"Clariss, Rajum"* [the most famous Ray] as the very best of his predecessors.) Ray, a man of humble background (his father was a blacksmith in the rural town of Black Notley in Essex), managed to parlay his intellectual gifts into a degree from Cambridge. For more than a decade he pursued his work in natural history under the patronage and partnership of a wealthy fellow Cantabrigian, Francis Willughby. The two friends traveled together throughout Europe during the mid-1660s, and then developed plans for a lifetime of work in publishing comprehensive joint volumes on the taxonomy of all organisms, plants as well as animals. But Willughby died in 1672, at age thirty-seven, and the disconsolate Ray, with steadfast loyalty and constant invocation of his friend's memory and intellectual contribution, soldiered on alone. Ray wrote nearly all of their two great joint volumes on birds (1676) and fishes (1686), but the funding, and a good share of the basic observation, represents the long hand of Willughby's generous legacy.

Their beautiful book on birds, illustrated with seventy-eight engraved plates (see figures 4 and 5 for an example of their art, and for the cover page, from my collection, of the 1678 English translation from the Latin original), begins with a preface, written by Ray, that restates, even more forcefully than Grew had done, but to the same effect and purpose, the embattled feeling of naturalists in these early years of the Scientific Revolution. Grew and Ray insisted that the literary and non-observational traditions of Renaissance humanism had built an

Figure 4.

intellectual barrier that must be breached before a properly empirical science of taxonomy and natural history could be formulated.

Ray begins by proclaiming the new and different methods that he and Willughby had rigorously followed. He then, and at greater length, forcefully criticizes the older practices that must now be abandoned. Above all, personal observation must replace ancient testimony as the primary ground of information:

Figure 5.

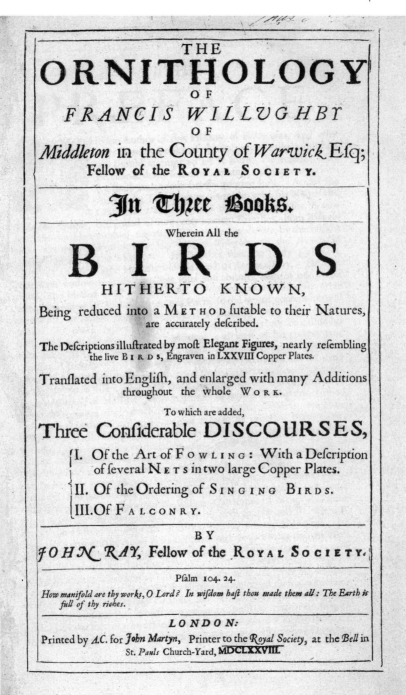

THE
ORNITHOLOGY
OF
FRANCIS WILLUGHBY
OF
Middleton in the County of *Warwick* Esq;
Fellow of the ROYAL SOCIETY.

In Three Books.

Wherein All the
BIRDS
HITHERTO KNOWN,

Being reduced into a METHOD futable to their Natures,
are accurately defcribed.

The Defcriptions illuftrated by moft **Elegant Figures**, nearly refembling
the live BIRDS, Engraven in LXXVIII Copper Plates.

Tranflated into Englifh, and enlarged with many Additions
throughout the whole WORK.

To which are added,
Three Confiderable DISCOURSES,

I. Of the Art of FOWLING: With a Defcription
of feveral NETS in two large Copper Plates.

II. Of the Ordering of SINGING BIRDS.

III. Of FALCONRY.

BY
JOHN RAY, Fellow of the ROYAL SOCIETY.

Pfalm 104. 24.

*How manifold are thy works, O Lord? In wifdom haft thou made them all: The Earth is
full of thy riches.*

LONDON:
Printed by *A.C.* for *John Martyn*, Printer to the *Royal Society*, at the *Bell* in
St. *Pauls* Church-Yard, MDCLXXVIII.

> We did not, as some before us have done, only transcribe other men's descriptions, but we ourselves did carefully describe each bird from the view and inspection of it lying before us.

Again, Gesner and Aldrovandi symbolize the bad old way of Renaissance compilation, and their techniques of promiscuous reporting and description, based primarily upon other writings and with special dignity accorded to Greek and Roman sources, must now be replaced:

> As for the scope and design of this undertaking, it was neither the author's [that is, Willughby's], nor is it my intention to write pandects of birds, which should comprise whatever had been before written of them by others, whether true, false or dubious— that having already been abundantly performed by Gesner and Aldrovandus, nor to contract and epitomize their large and bulky volumes, lest we should tempt students to gratify their sloth.

In his strongest statement, Ray contrasts the geometrically opposite patterns of Renaissance compendia and modern scientific treatises. Gesner and Aldrovandi, by including everything and imposing no selective criterion of veracity, compiled works of constantly increasing size and scope. He and Willughby, by insisting upon factual accuracy, preferably validated by their own eyes, would employ the opposite standard of whittling and excision, distinguishing the true from the false, the relevant from the incidental, and then publishing only the bare bones of dependable factuality. Ray also explicitly identified the omitted material as part of "humane learning" and endowed the good stuff, now separated and retained, with the honorable name of "natural history":

> We shall further add that we have wholly omitted what we find in other authors concerning homonymous and synonymous words, or the divers names of birds, hieroglyphics, emblems, fables, presages, or ought else appertaining to divinity, ethics, grammar, or any sort of humane learning; and present him [the reader] only with what properly relates to natural history. Neither have we scraped together whatever of this nature is anywhere extant, but have used choice, and inserted only such particulars as ourselves

can warrant upon our own knowledge and experience, or whereof we have assurance by the testimony of good authors, or sufficient witnesses.

Finally, in a statement ending with a first sign of forthcoming troubles that have grown with the centuries and have led to the circumstances surrounding the composition of this book and others of its genre, Ray explicitly brands as unworthy of much scientific attention the chief Renaissance goal of trying to link each modern bird with its ancient name. I raise no objection to this judgment, but Ray then appends a slightly gratuitous rebuke to the humanists—imbued with more than a whiff of the Philistine—ridiculing their undue attention to literary style, whereas good scientific prose only requires clarity, and need not fret about quality:*

> For to what purpose is it eternally to wrangle about things, which certainly to determine is either absolutely impossible, or next door to it? Especially seeing if by immense labor it might at last be found out by what names every species was known to the Ancients, the advantages that would thence accrue would not countervail the pains. About the phrase and style we were not very solicitous, taking greater care to render the sense perspicuous than the language ornate.

---

*Unfortunately, too much of modern science, although now so long, and so clearly, beyond any need thus to mark its territory of distinction from humanistic study, has retained and intensified this attitude into an active distaste for stylistic felicity in writing—as if the factual content of a work becomes debased if an author also possesses a fortunate talent for decent prose (a true perversion of the hedgehog's claim for one great way—a doctrine never intended to restrict synergistic approaches to the same *good* end—the hedgehog's true goal). Thus we may identify in Ray's last line a source of later trouble, a claim that would subsequently be read as arrogance and parochialism once the tables turned. I must also add the ironic observation that, despite his expressed judgment (or perhaps, one ought to say, in spite thereof), Ray happened to be an excellent writer, and his good prose certainly furthered his purposes.

4

The Mandate
of Magister Medice:
The Threat of Suppression

IN CONTROVERTING THE RENAISSANCE PROGRAM FOR RECOVERING THE wisdom of Antiquity, rather than winning novel insights and explanations by observation and experiment, the initiators of the Scientific Revolution worked to clear away the passive impedimenta of old beliefs. Breaking through the inertia of ages is no easy task, for incumbency brings enormous advantages both in politics and intellectual life. But active suppression poses far more serious problems, including actual danger to life and limb—and the avatars of the Scientific Revolution also faced (or at least often thought they faced, leading to a psychological burden that should not be underestimated, whatever the actual danger) a more than merely hypothetical threat of suppression or mayhem from reigning secular powers of the time.

Our simplistic renderings of Western history, as previously mentioned, tend to depict any struggle between science and secular power as part of a "warfare between science and theology," or as "science versus religion"—but I strongly reject this harmful and simplistic dichotomy (see pages 85–89 for more detail on this false model of history). Secular or state power did, at least in a few crucial episodes, actively suppress the spread of scientific methods and conclusions. Given the entanglements among major institutions at the time, the ideological basis for squelching a scientific claim usually found expression in religious terms—with arguments condemned (as in Galileo's canonical case) because they supposedly violated religious precepts that struck secular leaders as important in justifying their continued right to hold the reins of power.

In chapter 2 I reproduced a parody of a Catholic imprimatur, with the vice president of the Royal Society standing in for the official censor—a secular "blessing" for John Woodward's naturalistic effort in the physics of world-making in 1695. Figure 6 shows another example, also passed by the Royal Society, but this time preceding Grew's commissioned catalog of their collections, as dis-

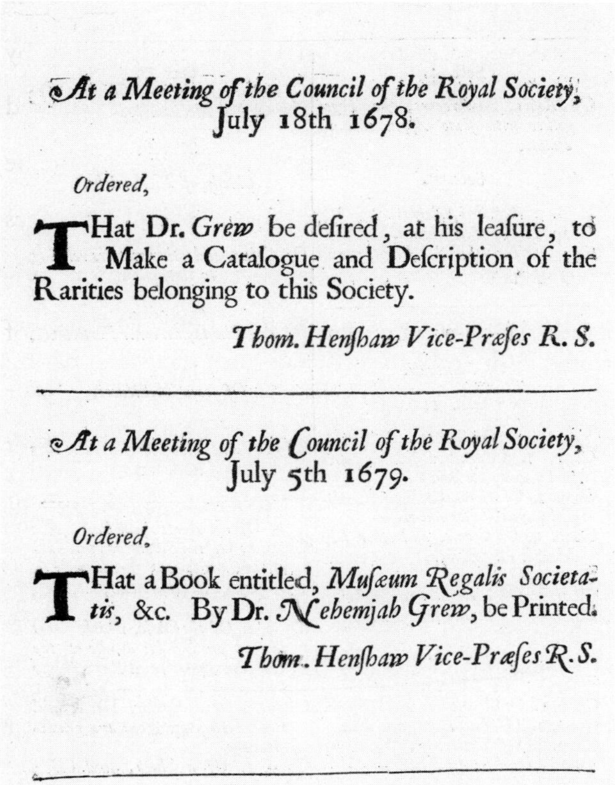

Figure 6.

At a Meeting of the Council of the Royal Society,
July 18th 1678.

Ordered,

THat Dr. *Grew* be defired, at his leafure, to Make a Catalogue and Defcription of the Rarities belonging to this Society.

Thom. Henfhaw Vice-Præfes R. S.

At a Meeting of the Council of the Royal Society,
July 5th 1679.

Ordered,

THat a Book entitled, *Mufæum Regalis Societatis*, &c. By Dr. *Nehemjah Grew*, be Printed.

Thom. Henfhaw Vice-Præfes R. S.

HOc opus Excellentiſſimi, & Celeberrimi Vlyſſis Aldrouandi, Patritij Bonó-
nienſis de Quadrupedibus ſolidipedibus volumen integrum perlegi diligen-
ter Ego Don Marcellus Baldaſſinus Clericus Reg. Congreg. S. Pauli, pro Archie-
piſcop. Curia Bononienſi reuiſor deputatus, & quod nihil contra ſanctæ fidei dogmata, vel probatos mores, aut ſacri Indicis regulas contineret, vt typis mandaretur
probaui.

Idem qui ſupra Don Marcellus Baldaſſinus.

EGo Fr. Hieronymus Onuphrius Romanus, ex Conuentu S. Mariæ Gratiarum
Doctor Collegiatus, & Lector publicus, ac Sanctiſſ. Inquiſitionis Conſultor,
mira quadam animi oblectatione, atq; attentione totum hoc aureum opus, ac ſatis
copioſum, inſcriptum de Quadrupedibus ſolidipedibus, & conſcriptū ab Illuſtriſſ.
atq; Excellentiſſ. Viro Vlyſſe Aldrouando Patritio Bonon. perlegi; cumque in eo
nihil repererim, quod aut pias piorum hominum aures offendat, aut ſit contra Ec-
cleſiaſticas regulas, ac ſanctiones, quin potius multum vtilitatis inde toti homini
emergere cognouerim, ideo in Dei gloriam, ac communem omnium vtilitatem
typis excuſſum in lucem prodire cenſui.

Imprimatur igitur
Idem qui ſupra Fr. Hieronymus &c. nomine Reuerendiſſ. P. Mag. Pauli de Garex.
Inquiſit. Bonon.

Figure 7.

cussed in chapter 3. To show the real McCoy, I now depict (figure 7) a genuine
Catholic imprimatur, as printed in an important Renaissance treatise in natural
history by a key figure in this book's preceding chapter (figure 8, from my copy
of Ulisse Aldrovandi's 1639 edition of his first volume on mammals, *De quadrupedibus solidipedibus* [four-footed beasts without cloven hooves], printed
posthumously, and including chapters on horses, unicorns, rhinoceroses, and
elephants).

The wording of the imprimatur, passed by two readers and then approved
by the Inquisitor of Bologna, strikes our modern ears as more than a bit chilling. The first censor approves, citing the conventional claim that he found
therein *nihil contra sanctae fidei dogmata, vel probatos mores* (nothing against
the dogmas of the sacred faith or accepted morals). The second reader, a bit

Figure 8.

more florid in his approval, finds nothing offensive either to the ears of pious people or to the rules of the Church. The mandate then proclaims *imprimatur igitur* (therefore let it be printed).

I don't mean to exaggerate the chilling effect of such pronouncements. All books published under Catholic auspices at the time had to receive this official sanction. I doubt that such a famous and uncontroversial character as Aldrovandi waited with bated breath; and the printed permissions, however contrary to modern ethics and sensibilities, tend to be formulaic and repeated from book to book, thus representing the boilerplate of their age—somewhat akin, perhaps, to packages of food passed by the Pennsylvania Department of Agriculture, or those old mattress tags that seemed to threaten death to any intrepid tamperer.

But I chose this particular example of a genuine imprimatur for a quite different, if adjacent, reason. If we flip the page of the imprimatur, we find on the other side, reproduced in figure 9, a symbolically chilling statement that sent a shiver up my spine because I wasn't expecting this further reminder of the true dangers of suppression, potentially including incarceration and bodily harm. Aldrovandi's dedication, printed on this page, reads: *Maffaei Card. Barberini nunc Urbani VIII Pont. Max.*—to Cardinal Maffeo Barbarini, now Pope Urban VIII. (I recently saw a copy of this volume's first edition, published a few years earlier, and before Urban's promotion. The dedication then included only the first line of type, praising the cardinal, not yet elevated.) Catholic intellectuals invested great hope in Maffeo Barbarini, an apparent friend of science, and of liberal learning in general. Galileo himself called Urban's election to the papacy a *mirabel congiuntura* (great conjuncture) that would strongly foster the respect and approval of science. But the same Urban VIII, ten years later, in 1633, sponsored Galileo's trial before the Roman Inquisition, and forced his recantation (followed by a life of house arrest) for daring to advocate the heresy of a central sun!

Since the desired alternative of respect and independence does not seem generally available as a realistic option, perhaps intellectuals should wear the suspicion or opposition of secular powers as a badge of honor. At least they seem to fear us (or, at minimum, to regard us as worthy of monitoring), even though our actual weapons rarely extend beyond the pen, or its recent reconfiguration as the electronic keyboard. This second legitimate fear that science felt in its infancy (and that has not become entirely extinct today, even with science in powerful maturity)—suppression by the politics of secular power, usually stated in overtly religious (in the past) or moral (today) terms—often reaches well beyond any explicitly scientific content in the work under scrutiny.*

As an example of this realistic unease among scientists, and of the reach (well beyond scientific content) that public suspicion can attain, I present a

Figure 9.

---

*Sometimes, to be fair, the justice of this scrutiny cannot be gainsaid because scientists, particularly after winning power and authority as members of a central and established institution, have often ventured beyond their sources of genuine expertise and claimed special insight into ethical issues for the logically invalid reason of superior factual knowledge about questions relevant to the debate at hand. (My technical knowledge of the genetics of cloning gives me no right, or expertise, to dictate legal or moral decisions about the politics, sociology, or ethics of creating, say, a genetic Xerox copy of a grieving couple's dead child.) But I speak, in this chapter, of legitimate claims by scientists for protection of intellectual work in their own magisterium of nature's factuality and causal operation.

striking example of turning tables (admittedly of a superseded sixteenth-century sort, but with a reminder that similar activities follow more subtle pathways in our time). In chapter 3, I illustrated the passive impediments that early scientists of Newton's generation felt from the traditions of Renaissance humanism—as represented in natural history by the compendiary tradition of Gesner and Aldrovandi. But, to show that these men also experienced their own more active impediments, I have reproduced Aldrovandi's imprimatur and dedication to Galileo's oppressor. I now present an example from his intellectual partner, Konrad Gesner, not based on Gesner's scientific content but on the intellectually irrelevant circumstance (at least for this book on four-footed beasts) of his Protestant affiliations, as the godson and protégé of the important Swiss reformer Ulrich Zwingli.

Many years ago I saved my pennies and finally purchased a copy of Gesner's first and greatest zoological work, volume one, titled *De quadrupedis viviparis* (on four-footed, live-bearing beasts—that is, terrestrial mammals in modern terms) of his *Historia animalium,* published in 1551. But when I looked at the title page (reproduced in figure 10), I encountered a fascinating puzzle, resolved only years later when I learned enough Latin to work through the rationales. This book, by the way, served as the prod, initiated long ago and building for more than a decade, to my decision to write this little volume—so I thank Gesner's expurgator for a personally rewarding spinoff from his dubious practices.

The peculiarities of this expurgation, described just below, have weighed on both my heart and mind ever since this purchase. I used the example, schlepping the book all the way to Washington for penance and prospect, as the focus for my millennial presidential address to the annual meeting of the American Association for the Advancement of Science in 2000—a disquisition on the relationships between science and humanistic study. I also, from this prod, conceived the admittedly quirky and idiosyncratic idea for this book—to base a volume on the well-trodden subject of science and humanities upon largely unknown examples taken from specific passages in antiquarian books from my own collection—a classical technique of humanist scholars, but now attempted by this card-carrying scientist! I, at least, have always felt that evidence of actual and original sources, right before one's own eyes, indeed in one's own hand, packs an almost indefinable and quite special emotional punch in its authenticity, at least in my reactions. I will never forget my grandmother's frequent remark that she would only credit something

Conradi Gesneri.

CONRADI GESNERI

medici Tigurini Historiæ Anima
lium Lib. I. de Quadru
pedibus uiuiparis.

OPVS Philosophis, Medicis, Grammaticis, Philologis, Poëtis, & omnibus
rerum linguarumǭ uariarum studiosis, utilissimum si-
mul iucundissimumǭ futurum.

AD LECTOREM.

HABEBIS in hoc Volumine, optime Lector, non solum simplicem animalium historiam, sed etiam ueluti com-
mentarios copiosos, & castigationes plurimas in ueterum ac recentiorum de animalibus scripta quæ uidere hactenus nobis
licuit omnia: præcipuè uerò in Aristotelis, Plinij, Aeliani, Oppiani, authorum rei rusticæ, Alberti Magni, &c. de animalibus
lucubrationes.   Tuum erit, candide Lector, diligentissimum & laboriosissimum Opus, quod non minori tempore quàm
quidam de elephantis fabulantur, conceptum efformatumǭ: nobis, diuino auxilio nunc tandem in lucem ædimus, non modo
boni consulere, sed etiam tantis conatibus (ut alterum quoǭ Tomum citius & alacrius absoluamus) ex animo fauere ac be-
nè precari: & Domino Deo bonorum omnium authori seruatoriǭ, qui tot tantasǭ res ad Vniuersi ornatum, & uarios hominum usus creauit, ac nobis ut ea contemplaremur uitam, ualetudinem, otium & ingenium donauit, gratias agere maximas.

TIGVRI APVD CHRIST. FROSCHOVERVM,
ANNO M. D. LI.

Figure 10.

when the original evidence stood right before her eyes *in schvartz*—that is, "in black" of print on paper.

So I gazed upon the title page of my proud acquisition and could not make out what I clearly saw. I could read Gesner's book title, and the last two words of his identification: *medici Tigurini* (doctor of Zurich). But his printed name had been obliterated in two distinct ways: first by a clever inking through the printed letters to create the gobbledygook of a meaningless string from the original name; and, second, by the further and literal cover-up (perhaps following the censor's dissatisfaction with his initial effort) of a strip of paper, once glued directly over the name, but later removed. (A subsequent owner, further countering the censor's earlier work, then reinserted Gesner's name, in ink and above the original blottings.)

This laborious, if almost whimsical, excision of Gesner's name continues throughout the book of 1,104 pages. Just consider (figure 11) the beginning page of text, where Gesner's name has been inked over and extended into the meaningless and unbroken string of letters, LOQNRIADIVOESNERIATI, just above the charming picture that accompanies the first chapter *De alce* (on elks) in his alphabetical accounting. As I then proceeded, page by seemingly endless page, through the rest of the book, the resulting pattern eventually struck me as utterly ludicrous rather than seriously evil. The Catholic censor who gained control of this copy faced a peculiar sort of problem: the book itself contained nothing objectionable in religious or moral terms. Gesner had simply recorded everything ever said about a bunch of mammals, and the *defensores fidei* found nothing offensive in principle therein. In fact, and on matters strictly religious, the censor did little beyond sprucing up the few biblical quotations that Gesner had cited from Luther's translation by laboriously appending the approved Catholic version from the Latin Vulgate (figure 12). And if these very minor differences of a word or two here and there, primarily in God's famous oration to Job from the whirlwind, have any theological significance, I can only say that the subtle distinctions thoroughly elude me.

Gesner's book had not been placed on the Index, and his more than a thousand pages contained enormous value to any Catholic reader interested in natural history. Yet the censor went through every last page, making funny little blottings here and there, *but very carefully*, throughout. What had he accomplished? When I finally recognized the pattern, I became more amused than offended. Gesner's words threatened no Protestant dangers, but his persona, and that of several other folks he quoted, did raise Catholic hackles, especially

LOQNRIA DIVGESNERIATI·
GVRINI HISTORIAE ANIMALIVM
LIBER I. DE QVADRVPEDIBVS
VIVIPARIS.

DE ALCE.

Picturam hanc à pictore quodam accepi, quam ueram esse testantur oculati testes: ut etiam cornua, quæ gemina habet. Nos unum hic seorsim pinximus.

L C E S, alcis: uel alce, alces: ἄλκη, paroxytonum potius quàm oxytonum: Germa-
nicè Elch uel Ellend, aliqui l geminant, alij aspirationem præponunt, ut apud
Latinos etiam nomnulli, quod non probo, Illyricè Los, Polonicè similiter, & apud
alios Pouuod, ut ex indigena quodam nuper accepi. Illyrij etiam ceruum Gelen
uocant, & fieri potest ut inde nomen huius animantis ad Germanos translatũ sit,
propter similitudiinem eius cum genere ceruino. Nullum huius animalis nomen
aliæ gentes habent, cum peregrinum omnibus sit præterquam Scandinauiæ, quod sciam: proinde
non assentior Iudæis illis, qui Deuteronomij cap. 14. זאמר zamer alcen interpretatur: quãquam alij
pro eadem rupicapram, alij camelopardalin reddunt: mihi ad postremam animus magis inclinat.
In tam rara igitur & longinqui soli fera authores inter se uariare, minus mirabimur. Ego singulo-
rum uerba apponam seorsim, cum alioqui satis commode conciliari non possint. Inter Græcos so-
lus Pausanias (qui Antonini tempore claruit) in Eliacis differens de elephanti uulgo creditis den-
tibus, quod cornua sint nõ dentes, haud omnibus enim eodem loco cornua nasci: argumento sunt,
inquit, Aethiopici tauri, & alcæ feræ Celticæ, ex quibus mares cornua in superciliis habent, foe-
mina caret. Sed forte hoc loco Pausanias alcen confundit cum quadrupede illa quam hodie ran-
giferum uocant, cui cornu è media fronte procedit, ut suo loco dicemus. Eiusdem in Bœoticis uerba hæc sunt: Alce nominata fera, specie inter ceruum & camelum est, nascitur apud Celtas, explo-
rari inuestigariç ab hominibus animalium sola non potest: sed obiter aliquando, dum alias uenan-
tur feras, hæc etiam incidit. Sagacissimam esse aiunt, & hominis odore per longinquum interual-
lum percepto, in foueas & profundissimos specus sese abdere. Venatores montem uel campum ad
mille stadia circundant, & contracto subinde ambitu, nisi intra illum fera delitescat, non alia ratio
ne cam capere possunt. Hæc Pausanias, qui ut plerisç ueteres Germaniam totam & Septentriona-
les finitimas regiones uno Celticæ momine comprehendit.

¶ Cæsar lib. 6. Commentariorum de bello Gallico: Sunt item in Hercynia sylua quæ appellan-
tur Alces, harum est consimilis capris figura, & uarietas pellium: sed magnitudine paulo antece-
dunt, mutilæç sunt cornibus: & crura sine nodis articulisç habet: neç quietis causa procumbunt:

a

Figure 11.

DE CVRA ET PROVIDENTIA QVA
DEVS BESTIAS RATIONIS EXPERTES
dignatur & prosequitur, locus lectu dignissi-
mus ex Iobi capitibus 38. & 39.

NVm tu uenaberis L E O N I prædam, & pastum catulis eius suppeditabis? Quis C O R V O de cibo prospicit, cum pulli eius famelici ad Deum clamantes oberrant? Nostine tempus quo C A P R A E F E R A E in rupibus pariunt? An obseruasti partum C E R- V A R V M, aut numerum mensium quos implent, & parturiendi tempus? Submittunt se illæ, in- curuatæ̃q́ fœtum magnis doloribus ædunt. Tum hinnuli adolescunt, & pabulo iam confirmati re- licta matre non redeunt. Quis A S I N V M syluestrem (pere) liberum dimisit, aut quis O N A G R I (arud, Hebræi tum pere tum arud asinum ferum interpretantur) uincula soluit? Ego domicilium eius in solitudine po- sui, & cubile in loco sterili. Itaq́ ridet turbam oppidanam, nec audit clamores agasonis. Pascua sibi in montibus disquirit, & stirpes omne genus uirentes sectatur. Voletne M O N O C E R O S tibi seruire, aut morari ad præsepe tuũ? An loro ipsum uincies, ut sequendo te sulcos aratro imprimere, aut glebas frangere uelit? Ausisne illi credere, tantõq́ robore præstanti tuum permittere laborem? Sperabisne messem tuam ab eo conuehendã, ut condatur in horreum? En pulcherrimas S T R V- T H I O N I S (pauonis secundum alios, aut galli syluestris) alas, quantũ superant pennas, & alas C I C O N I A E? (Quidam uertit: An [dedisti] alas plausibiles pauonibus, aut pennas ciconiæ & plumas?) Sed deserit in terra oua sua, ut in pul- uere foueantur: nec cogitat pedibus ea dissipari, & à bestijs conculcari posse. Ita immitis est in pullos suos, ac si sui non essent: & ita pro eis solicita nõ est, ut peperisse frustra uideatur. Nullam enim men- tem aut intellectum diuinitus accepit. Quo tempore uero sublimis euolat, ridet equũ simul & equi- tem. Túne E Q V O dabis ut generosus & bellator sit: ut alta ceruice ferociat, & hinnitum ædat? An speres te illum excitare aut terrere posse instar locustæ? Atqui nares eius ferociam spirant: calci- bus solum fodicat, & fortitudine sua superbus armatis occurrit. Metum omnem contemnit, nõ fran- gitur animo, non expauescit micantem gladium. Non pharetræ sonitum, non hastam uibratam, non lanceam aut cuspidem curat. Dumq́ fremitus & tumultus cietur, terram fodit, nec tubæ sono mo- uetur. Classico tubæ signo animosé adhinnit, ac eminus prælium, & ducum clamorem tumultumq́ tanquam odorans percipit. Num per tuam sapientiam fit, ut A C C I P I T E R uolans alas suas uentis committat? Tuóne iussu sublimis A Q V I L A fertur, & nidum in alto struit? Incolit illa (Vultur incolit petras, & c. LXX. Sed cadaueribus pasci Matthæi etiam cap. 14. aquilæ non uulturi adscribitur) petras, & inaccessas rupium ueluti arces: inde sibi de esca prouidet, longe latéq́ perspicacissima circumspectans. Pulli eius sanguinem sorbent: ipsa cadaueribus ubicunq́ fuerint, aduolat.

DE FRVCTV EX ANIMALIVM HISTORIA
PERCIPIENDO, EX THEODORI GAZAE
præfatione in conuersionem suam Aristotelis
de animalibus librorum.

Physicus quo-
modo uersetur
in historia ani-
malium, & in
quem finem.

OMnis philosophandi ratio naturalis, ubi à primis illis naturæ initijs, materiam dico, formam, finem, agens, & motum (ut ita loquar) emerserit, hic uersatur, ac diutissime immoratur, hic suas uires exercet, atq́ multiplicem, uariam, & admirabilem rerum constitutionem amplissime explicat. Persequitur deinde discrimina omnia, quibus natura suas animantes differre inter se uoluit: colligit summa genera, reliqua sigillatim exponit: partitur in species genera: & singula, quæ circiter quin- genta numero in his continentur libris, describit: pergit quæq́ explanans, quemadmodum oriãtur, siue terrestria, siue aquatica: quibus nam constent membris, quibus uescantur alimentis, quibus affi- ciantur rebus, quibus moribus prædita sint, quantum uiuendi spacium datum cuiq́ est, quanta cor- poris magnitudo, quod maximum, quod minimum est: quæ forma, quis color, quæ uox, quæ inge- nia, quæ officia: deniq́ nihil omittit, quod in animalium genere natura gignat, alat, augeat, & tuea- tur. Quæ omnia eõ spectant, ut, quod sanctissimus quoq́ author ille, quem deus sibi ueluti suppelle- ctilem quandam preciosam elegerat, admonet, ex ijs, quæ à natura proueniunt, deum immortalem, ex quo ipsa pendet natura, intelligamus, admiremur, atq́ colamus: qua re nihil pulchrius, nihil gra- uius, nihil dignius homini esse potest. Tantus fructus horum librorum est. Nec audiendi sunt, qui inquiũt: Multa Aristoteles de musca, de apicula, de uermiculo, pauca de deo. Permulta enim de deo is tractat, qui doctrina rerũ conditarũ exquisitissima, conditorẽ ipsum declarat: nec uero musca, nec uermiculus omittẽdus est, ubi de naturæ mira solertia agitur. Vt enim artificis cuiusuis, sic naturæ in- genium in minutissimis potius contemplandũ est. Quinetiam cum r̃ erũ causas cognoscere pul- cherrimũ sit (hac enim una cognitione, homo perfici, absoluíq́ potest, ut deo immortali similis, quo- ad eius fieri potest, euadat) his sané libris plene docemur, cur quæq́ res in animalium genere ita sit: planéq́ felicitatem assequimur illam nobiliorem, quæ in actione animi consistit, quam sapiens quoq́ poëta

Minutorum ani-
malium contem-
platio non sper-
nenda.

Causarũ cogni-
tio quàm no-
bilis.

in this raw first generation, just after Luther's heresies shook Catholic complacency to the roots and launched a vigorous response, known as the Counter-Reformation. Thus the censor did very little more, despite extensive effort and expenditure of time, than blot out a few objectionable names, whenever Gesner had dared cite them in print. Moreover, more than half the erasures simply obliterate two names that run throughout Gesner's text for obvious reasons. (Both men, in fact, remained Catholic, but their iconoclasm and lack of pious orthodoxy made them personae non grata in these highly fractious times.) First, the great Erasmus of Rotterdam (1469–1536), perhaps the most celebrated scholar of the Renaissance, who wrote virtually nothing about zoology as we understand the subject today, but who compiled, in his *Adagia,* the most complete book of proverbs ever assembled (see my preface, page 2). Since Gesner's Renaissance compendium cites everything ever said about mammals, with stress on human concepts of their natures and powers, he includes an explicit section on proverbs in each of his chapters, prominently listing all of Erasmus's entries, properly attributed. So the censor laboriously excised every mention of Erasmus's person, and retained all the words about the animals themselves. (Well . . . perhaps not *every* last one. So do stay tuned, for our censor, you shall soon learn, is the Magister of this book's title and Erasmus is the source of our leading motto about the fox and the hedgehog. So all these themes may just come around again, dear reader, if you maintain the patience to persist to this book's end, where all these creatures will coalesce to make a closing appearance to conclude this volume on a point of hope!)

Second, Sebastian Münster (1489–1552), whose *Cosmographia* of 1552 described the geography and biology of all parts of the known world—another obvious source for Gesner's copious citation. At least the censor permitted himself some fun in an otherwise tedious task, for he followed different and consistent schemes in blotting out the various names (figure 13). Erasmus merited just a thick black band through his entire name, but Münster enjoyed enough filigrees over, around, and through his letters to make his moniker unreadable.

Gesner includes a bibliography in the front matter of his book, so we can quickly discern the pattern of names chosen for excision—a pretty simple matter of Protestant (or renegade Catholic) bad, orthodox Catholic good. For example (see figure 14), Christopher Columbus, at Number 169, comes up golden in claiming the New World for the Catholic Majesties of Spain. But

Figure 12.

442 De Quadrupedibus

μυίας ποιῶν, Eraſmus ex Suida:meminit etiam Apoſtolius. Nihil ab elephante diſſe[r]
φόρτας ὀσλῶ, in magnos & ſtupidos dicebatur:etiamſi primā ingenij laudem Plinius m[ ]
ſed inter bruta. Verum corporis moles & formæ fœditas, adagio locum fecit. R[ ]
no. Videtur huc alludere Palæſtrio Plautinus, qui herum ſuum non ſuo, ſed eleph[ ]
teċtum ait, nec plus habere ſapientiæ quàm lapidem, **Budæus**. Legitur etiam ap[ ]
ſtolium, ὐδὶ τῶ μεγάλωψ καὶ ἀναιδ̇ήτωψ, πάρόσιψ καὶ τὸ ζῶοψ ποιõνϣ. Apud Epinicum in [ ]
cum iaċtaret quidam ſe elephantum poculum tricongium, quod ne elephas quidem [ ]
exiccaſſe, ſubijcit quidam, Ὀυδὲψ ἐλέφαψτ⊙ γὰρ ǒἰαφέρεις ὀδὲ ſυ, ut ſupra ex Athenæo [ ]
uer ſibus poſtea obſeruatis **Budæus** etiam hæc uerba, Ὀυδ̇ ἂψ ἐλέφας ἰκπίοι, id eſt, N[ ]
ebiberet, adagijs inſeruit. Rhytus (inquit:errat autem, nam pro poculo rhytum, [ ]
ſemper profertur) poculi genus eſt, ſpecie cornu, quod uidetur (hoc ex ſua cõieċtu[ ]
dem non uidetur) eburneum fuiſſe, impoſitum imagini elephanti:ita ut quadruple[ ]
elephanti mentio fiat (poculi, imaginis belura, eiuſdem uiuæ, & hominis ebibentis, [ ]
ti nomine ὁμωνύμως diċtorum.) Dicetur, inquit **Budæus**, in librum inſulſum ac loqu[ ]
patientiſſimus quidem perlegere ſuſtineat. Magnos ſtupidoſꝗ elephantorum [ ]
bant, uel Græco ſuffragante prouerbio, Cælius. Celerius elephanti pariũt: Sunt q[ ]
(inquit **Budæus**) inter adagia uidetur adnumerãdum, quod ſcriptum eſt apud Plin[ ]
in præfatione hiſtoriæ mundi:Nam de Grammaticis, inquit, ſemper expeċtaui par[ ]
bellos meos, quos de grammatica ædidi:& ſubinde abortus fecère iam decem annis, [ ]
tiam elephanti pariant, Haċtenus Plinius. Itaꝗ cunċtationem immodicam, & qu[ ]
ta molimina, his uerbis licebit ſignificare. Porrò de elephantorum partu Plautus in [ ]
pe hoc uulgò dicier, ſolere elephantum grauidam perpetuos decem eſſe annos. Lic[ ]
in hāc uertere formam:Quando tãdem paries obſecro, quod tot iam annos uterum ge[ ]
ti diutius:De elephantorum pariendi tempore ſententias authorum diuerſas, capite [ ]
eburna uagina plumbeus gladius, ἐψ ἐλεφαψτίνῳ κελ̄ᾳ τὸ μολύ βδιψοψ ξίφ⊙: prouerbium [ ]
genis Cynici apophthegmate, Nam cum adoleſcēs quiſpiam inſigni forma, fœdum q[ ]
ſcœnum dixiſſet:Ex eburna, inquit, uagina plumbeum gladium educis. Ebur [ ]
facere, eſt genuinæ formæ, cultum atꝗ ornatum externum inducere, quo decus illud[ ]
retur magis quàm illuſtretur. Proinde læna Plautina puellæ naturali forma prædita, [ ]
ad oblinendas malas poſtulanti: Vna, inquit, opera ebur atramento candefacere po[ ]

DE EQVO.

A.

Q V V S nobiliſſimum inter quadrupedes animal, & uitæ humanæ mult[ ]
dis utiliſſimum, è iumentorum numero cenſetur:cui equidem nullum a[ ]
ter bruta exiſtimo, cum ingenij ſimul & corporis eius dotes perpendo. [ ]
quadrupedum ſuas laudes, & nonnullas quibus equum fortaſſis excel[ ]
boue pluribus modis uicium iuuet humanum animal nullum eſt. Sed ſi cõ ſeras inter f[ ]
licet uno aut altero equus uincatur, pluribus ſemper uincet. Accedit quod ubiꝗ terr[ ]
& naſci poteſt. Quamobrem merito prima ei quadrupedum, imò animalium omnium, [ ]
cipuè in planis regionibus, ut boui in montanis. Sed equi laudes & utilitates plurim[ ]
ſa hæc eius hiſtoria oſtendet, eò cæteris prolixior, quò plura de hoc animante, utpote [ ]
ſimo, apud authores inuenimus. ¶ Equum Latini etiam caballum uocant, de qua uoce [ ]
initio capitis oċtaui. Hebręi סוס ſus, ut equam ſuſah, quam uocem Canticorum p[ ]
tatum uel multitudinem equorum exponũt. Appellant autem Græci quoꝗ, tum equ[ ]
tum, hippon in fœminino genere:סוס quidem Hebraice equum aliqui diċtum pu[ ]
gaudio. Hieremiæ oċtauo ſus uel ſis auem quandam ſignificat, quam R. Salomon [ ]
ċtam, & Gallicè gruem uidetur interpretari: Sunt autem uerba prophetæ, de ijs a[ ]
turture, hirundine & grue, quæ norunt tempus ſuum migrandi ac redeundi. Sicut g[ ]
do ſic garriebam, meditabar ſicut columba, &c. Eſaiæ 38. pro grue Hebraice סוס ſus [ ]
equum exponit, ut etiam Ionathan, qui in iam citato Hieremiæ loco equum expoſu[ ]
equum reddit. Hieronymus alibi miluum, alibi hirundinis pullum. Septuaginta & Sy[ ]
σίωψ, id eſt hirundo. Εχοίας, equus Syris, Varinus. עברכש rekeſch, Dauid Kimhi docet [ ]
re equum præſtantem & non annis confeċtum ſic appellari: ipſe Kimhi aliud genus [ ]
ſcio quod (iumenti, פלנשטהשה) eſſe ſuſpicatur:& tertio Regum capite quarto ſcribit q[ ]
dim, id eſt, mulos interpretari. Leui ben Gerſon equos uelociſſimos intelligit, quorum [ ]
ſit;& ſimiliter author Concordant, genus equorum. Hieronymus Eſther oċtauo, com[ ]
Geneſeos 14. equitatum intelligit per רכש rekeſch, poſſeſſionem uerò per רכוש rekuſch [ ]
decimo rekuſch apparatum bellicum notat, alibi poſſeſſionem pecorum & rerum in[ ]
Achaſtranim Eſther oċtauo Hieronymus ueredarios transfert, Dauid Kimhi & A[ ]

Erasmus, at Number 171, disappears both for his *Opera* (that is, his collected works) and specifically, on the next line, for his *Adagia* (proverbs). At Number 178, Gaspar Heldelin (whoever he may be) gets the shaft for his *Ciconiae encominum* (his encomium on storks, whatever that may be). But the great Catholic geologist of Germany, Georgius Agricola, at Number 179, passes for his famous work on metals, weights, and measures, and also, on the next line, for his curious little pamphlet *De animantibus subterraneis* (on living things found underground, including a serious discussion of the gnomes that inhabited German mines, at least by the testimony of local laborers). But the Englishman William Turner, presumably an apostate and supporter of Henry VIII's takeover of monasteries, gets axed (in elegant filigree to obscure his name) at Number 183 for his book on birds.

It took me a while to realize—and more time to translate the tiny, if elegant, handwriting, replete with abbreviations—that the key to this peculiar form of "suppression lite" could be found in the cryptic line (figure 15) penned on the blank page just before the title (my thanks to David Freedberg and Tony Grafton for puzzling through this with me with their far superior knowledge both of Latin and of sixteenth-century handwriting):

> *Sine anathematis periculo liber iste d. historia animaliu. quadrupediu. viviparos legi potest. Na. ex mandato b. R. C. Inquisitionis Pisano diocosis Mag[ist]ri Lelii medices expuncta ac obliterata sunt ex albo quae del[en]da visa sunt.* (This dangerous book on the history of live-bearing, four-footed animals may be read without anathema. For, under the mandate of Magister [literally "teacher," but probably, in this case, meaning a graduate of the university] Lelio Medice of the Holy Roman Catholic Inquisition of the Diocese of Pisa, all [passages] that ought to be [so treated] have been expunged and obliterated from this volume.)

A bit chilling—what else can one say—despite the bumbling and innocuous character of the copious excisions. Magister Lelio Medice will not go down in history as a friend of science or scholarship—even though he has won some dubious form of transient notoriety in the title of this book!

And yet, before leaving this subject and closing the first part of this book,

Figure 13.

Catalogi

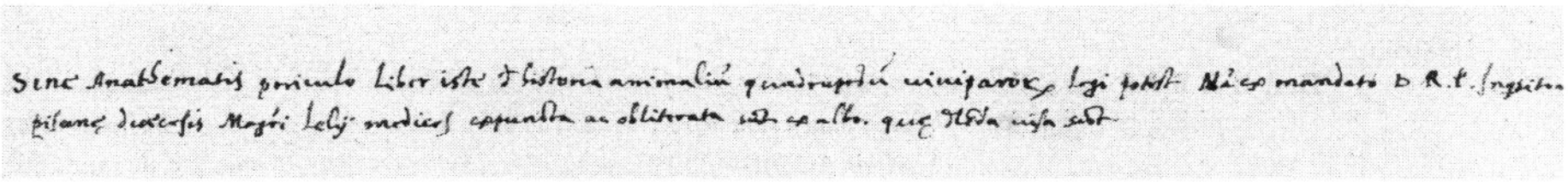

Sine Anathematis priuculo Liber iste d historia animaliū quadrupedū uiuiparoū, legi potest Nisi eo mandata D R.t. Inquisit. pisane Inuensis Magri Lelij medices expunata an oblitterata isd eo alto. quiq Ifola uisa sit

Figure 15.

I should state that I advance no secular conviction that bookburning, expurgation, and unpersoning represent exclusive strategies of religious dogmatists and other bedfellows of reactionary movements dedicated to protecting the status quo from all social or intellectual novelty. Our all-too-human temptation to censor or annihilate perceived enemies transcends the particularities of institutions, sacred or secular, and spreads across the full political spectrum, right to left. As a painful example—for this incident led to the death of one of the greatest scientists in history—consider the title page of an apparently humble pamphlet of enormous practical and historical importance, a booklet of instructions for the establishment of workshops to produce purer forms of saltpeter, an essential ingredient of gunpowder (figure 16).

This edition, published in 1793 at the height of ardor for the most radical phase of the French Revolution (including the Reign of Terror), reprinted a work that had been written and edited in the most auspicious of all years for revolution, 1776, and then published in 1777. The great chemist Antoine Laurent Lavoisier, appointed as *régisseur des poudres* (director of gunpowder), had perfected the manufacturing techniques, and then written most of the resulting pamphlet, that gave France the finest supply of purified gunpowder in the world. Indeed, without Lavoisier's successes, the beleaguered revolutionary armies might not have been able to repel the powerful invasion of otherwise better equipped and far more numerous foreign troops that had threatened to overthrow the new government.

This title page surely does not scrimp in expressing signs of revolutionary ardor, including the martial symbol of drums and banners, and the giveaway date printed at the bottom of the page: *"An II de la République, Une et Indivisible"*—year two of the Republic, one and indivisible. (The revolutionary government began time anew at the foundation of the Republic in September 1791, and then introduced a novel calendar of twelve months, named for weather and climate rather than monarchs and gods, each with

Figure 14.

INSTRUCTION

SUR L'ÉTABLISSEMENT

DES NITRIÈRES,

ET SUR LA FABRICATION

DU SALPÊTRE.

A PARIS,

Chez CUCHET, Libraire, rue & maison
Serpente.

AN II DE LA RÉPUBLIQUE, UNE ET INDIVISIBLE.

Figure 16.

thirty days, and with a celebratory addition of five days [six for leap years] at the end of each sequence.) However, amid this revolutionary paraphernalia, we must also note one conspicuous omission from the title page—the name of the author, the great Lavoisier himself. No mystery attends the excision of authorship, for at the moment of publication, during the Reign of Terror, Lavoisier languished in jail, under capital sentence for the supposedly non-capital offense of excessive diligence in his day job as a collector of taxes. Thus Lavoisier's name disappeared from the discoveries and publications that had saved a revolution now ready to terminate his life as well. Lavoisier had his date with the guillotine just three months before the abrupt termination of the Terror and the subsequent guillotining of the guillotiner Robespierre. The bitter eulogy of Lavoisier's dear friend, the mathematician Lagrange, may stand as a dramatic and more than merely symbolic reminder of how slowly we build our fragile intellectual structures, and how rapidly they can crumble when the zealots and philistines grab power: "It took them only an instant to cut off that head, but France may not produce another like it in a century."

II

FROM PARADOXICAL AGES OF BACON TO SWIFT SWEETNESS AND LIGHT

5

The Dynasty of Dichotomy

BACON'S PARADOX, NEWTON'S APHORISM, AND THE ADULT USE OF MOTHER GOOSE

IN PROMOTING THE CAUSE OF NEW KNOWLEDGE, WON BY OBSERVATION and experiment under a basically mechanical view of natural causation, and in denying the Renaissance's chief premise that scholarship would best advance by recovering the superior understanding achieved in Greece and Rome, the leaders of the Scientific Revolution popularized two metaphors with long pedigrees in Western literature. But these old sayings developed sharp edges in a quite conscious (and often virulently contentious) argument that swept through the intellectual world of seventeenth- and early-eighteenth-century France and England, and entered the record of history as the debate between Ancients and Moderns.

Francis Bacon, avatar of the Scientific Revolution, pushed his favored image so hard, and so often, that the saying became widely known as Bacon's paradox. The formulation is, indeed, a true and literal paradox—that is, a problem with two contradictory resolutions, each logical and correct in its own context. Bacon noted that our reverence for the classical giants of Greece and Rome often rested upon an impression of their venerable antiquity as expressed in their maximal distance (among known literary cultures) from our current efforts. At this great separation from ourselves, Plato and Aristotle seem old and full of wisdom. But, Bacon then observed, such an accounting might well be viewed as proceeding in exactly the wrong direction. After all, if knowledge accumulates through time, then with respect to a beginning point way back when, Plato can only be reckoned as a child and we must be deemed the old graybeards. For Plato and Aristotle cavorted during the youth of the world, and can only represent the bumptious boyhood of scholarship, while we Moderns carry both the accumulated weight of their youthful insight plus everything added since.

Bacon expressed this paradox in a famous aphorism: *Antiquitas saeculi, juventus mundi*—or roughly, the good old days were the world's youth. Why, then, should we indulge the Renaissance reverence for a time that can only represent the adolescence of knowledge, not the completion of wisdom. Time itself, not authority, Bacon added in a truly memorable line, is the "author of authors." Bacon, who knew and respected the Ancients even as he denied their claims to inherent superiority, then reminded his readers of the famous classical aphorism: "Truth is the daughter of time."

If Bacon was the avatar, Isaac Newton represents the apotheosis of this triumphant movement. We owe the second, and more famous, epigram (and visual icon) of the Scientific Revolution to a statement in a private letter that Newton wrote in February 1675 (as he dated the page, but as most of the rest of Europe, following the reformed Gregorian calendar, would have called 1676) to Robert Hooke, his colleague of similarly crusty disposition—a conjunction of temperament that led to frequent personal tension despite their basically similar view of life. In the midst of a personal squabble involving proper credit for work on the theory of colors, Newton, with uncharacteristic modesty and conciliation, wrote to Hooke, "If I have seen further, it is by standing on the shoulders of giants."

The two statements apply markedly different images to the same basic argument: the affirmation that knowledge progresses through time, and that the procedures advocated by the Scientific Revolution, rooted in observation and experiment under a mechanical view of causality, can best fertilize this growth, whereas the model of recovery, advocated by Renaissance scholars, must stymie progress by misreading an inchoate beginning as a completed acme. But Bacon's formulation is more pungent and unforgiving, while Newton's strikes a chord of diplomacy in affirming our reverence for the Ancients and asserting that we can exceed their achievement only because we add our puny novelties upon their magnificent foundations.

The history of this metaphor about the shoulders of giants, clearly devised to have it both ways (rendering obeisance to the Ancients while still asserting the accumulative character of knowledge and the consequent improvements of Modern times), enjoys a long and remarkable pedigree. (Most scientists credit the remark to Newton as a witty and original statement. Those who know differently often accuse Newton of sneaky borrowing, if not outright plagiarism, because he put no quotes around the utterance and cited no prior sources. But such claims are silly and trivial. After all, Newton wrote the line in a private letter to Hooke. He knew perfectly well, even if we have since forgotten, that he cited a standard image of his widely shared culture. Why would he invest the statement in quotation marks, or cite sources as if he were writing a scholarly paper? Do I, in every e-mail to a colleague, have to cite Andy Warhol if I talk about fifteen minutes of fame, or Churchill if I mention the end of the beginning?)

In fact, the pedigree of giant shoulders includes so much interest and weight of material that the great sociologist of science, Robert K. Merton, wrote one of the wittiest, yet deepest, works of modern scholarship by devoting an entire volume to pre-Newtonian uses of the image— *On the Shoulders of Giants* (New York, Free Press, 1965). Merton traces the depiction at least back to the twelfth-century lancet windows in the south transept of the Cathedral of Chartres, where the four Gospel writers of the New Testament appear as dwarfs sitting upon the shoulders of four great Old Testament prophets, Isaiah, Jeremiah, Ezekiel, and Daniel, depicted as giants. To show how rich, how picky, how contentious, how nuanced, and how subtle the history of this image can be, Merton devotes several learned chapters (of wonderfully light touch) to the seemingly endless wrangles among scholars about whether the Moderns who see farther by sitting atop Ancient shoulders must be depicted (as at Chartres) as dwarfs—so that the Renaissance conviction about superiority of the Ancients may be respected, even while we affirm the growth of knowledge—or whether the Moderns may be envisioned as equal in stature to the Ancients. (Some kind souls even objected that full-sized Moderns would surely strain the backs and bones of enfeebled Ancients, and that dwarfs must be preferred if only to spare poor Plato and Isaiah, thus making their yoke easier and lightening their literal burden.)

To show how far this wrangle could extend, and to quote from one of the most delightful documents of the time (a treatise that both Merton and I would love to rescue from its undeserved oblivion), George Hakewill, the Archdeacon of Surrey (and therefore not a practicing scientist, but a theologian who proves, thereby, that this seventeenth-century struggle did not pit science against religion) crafted a spirited defense of Modernist convictions in his prize essay, written for the official philosophical disputation at the Cambridge commencement of 1628. Hakewill (1578–1649) lit into the common and pessimistic belief that the entire universe, from the history of planets to the geography of landforms to the chronology of civilizations, marched inexorably into continuous decrepitude and decay—a process that must soon culminate in the earth's destruction. To the contrary, Hakewill argued, physical history has been stable, or quieting down from an initially distressing chaos, whereas the chronology of civilizations has featured continual progress in knowledge, morals, and sensibility, just as the Moderns argued against claims for the superiority of Ancient wisdom.

Following his generation's penchant for generous titles, Hakewill called his treatise *An Apologie or Declaration of the Power and Providence of God in the Government of the World. Consisting in an Examination and Censure of the Common Error Touching Nature's Perpetual and Universal Decay.* The book surely enjoyed some initial éclat. John Milton composed Latin hexameters for distribution during the Cambridge debate, and Samuel Pepys said of Hakewill's volume, "I fell to read a little in it, and did satisfy myself mighty fair in the saying that the world do not grow old at all."

Hakewill's spirited argument proceeds in a clear and persuasive order. He first dismisses all claims for physical decay either of the cosmos or the earth. He then takes up the strongest case for progress in human history: the accumulation of empirical knowledge about physical and organic phenomena—in other words, the improvement in what we would now call scientific understanding. Hakewill even dares to criticize the ultimate Greek and Roman standards of Aristotle and Pliny: "It is most certain that even Aristotle himself and Pliny were ignorant of many things, and wrote many not only uncertain, but now convinced of manifest error and absurdity."

Hakewill then moves to his most difficult task of arguing that manners and morals, and not only the more obviously accumulative character of purely factual information, have also improved through time, making modern Europe a paragon of rectitude in comparison with the supposed refinement of

Greek and Roman society. The titles of Hakewill's numerous chapters provide a good summary of both his general argument and the force of his prose. Hakewill particularly emphasizes Roman excess: "Of their long and often sitting and usual practice of vomiting, even among their women, as also of the number of their courses at a sitting, together with the rarity and costliness of their several services." "That their riot did not only show itself in the delicious choice of their fare, but in voracity and gormandizing, in regard of the quantity some of them devoured at a meal." "Of the Romans' excessive luxury in dressing and apparel. How effeminate they were in regard of their bodies, especially about their hair."

Hakewill's texts are endlessly entertaining, ranging from infanticide through human sacrifice to this description of the laws of Lycurgus, the traditional (but perhaps mythical) founder of the practices of Sparta in the seventh century B.C.:

> He ordained other laws so much in favor and furtherance of lust and all carnality, yea in the worst kind, that it might justly be said he made his whole commonwealth worse than a bordello. For he instituted certain wrestlings and dances, and other exercises of boys and wenches naked, to be done in public at divers times of the year, in the presence both of young and old men, which what effect it might work in the minds and manners of their citizens any man may easily judge.

But moving back to the shoulders of giants, Hakewill strongly affirms that we cannot ascribe any putative superiority of Ancient ways to an inherent decay in nature, but only to the bad, yet eminently correctable, habits of modern folks: "For matter of learning and knowledge if we come short of the Ancients, we need not impute it to nature's decay; our own riot, our idleness and negligence in regard of them, will sufficiently discharge nature, and justly cast the blame upon ourselves." Hakewill then quotes the sixteenth-century Spanish scholar Juan Luis Vives, strongly rejecting the polite and diplomatic tradition of depicting Moderns as dwarfs upon the shoulders of Ancient giants. We are all, Hakewill asserts, the same size, as he translates Vives's Latin statement into English (fifty years before Newton invoked the same image):

> For a false and fond similitude it is of some, which they take up as a most witty and proper one, that we being compared to the

Ancients, are as dwarfs, or they giants, but we are all of one stature, save that we are lifted up somewhat higher by their means, conditionally there be found in us the same studiousness, watchfulness and love of truth as was in them: which if they be wanting, then are we not dwarfs, nor set on the shoulders of giants, but men of a competent stature groveling on the earth.

I readily confess my chief aim in presenting the famous seventeenth-century quarrel of Ancients and Moderns as, at least in part, a birth pang of the Scientific Revolution, and a way of understanding an inevitable suspicion that arose at this time between nascent scientists and entrenched humanists, a mistrust that should have dissipated long ago, but has unfortunately persisted as our legacy today. That is, I wish to show the complexity and multifaceted character of this founding debate, so that we do not conceptualize the birth and later history of modern science as a war with two unambiguous sides, a clean dichotomy between dogmatic and hidebound humanists holding the fort of Antiquity against a progressive assault and inevitable breach by defenders of free inquiry and the power of new discovery. First of all, no mutual hatred ever existed; nearly all founders of the Scientific Revolution revered (and liberally quoted) the great sources of Antiquity. They also believed (and proved) that knowledge could progress by building upon those admirable foundations—the point of both Bacon's paradox and, particularly, Newton's admirable image of Antiquity as a firm foundation anchored by intellectual giants. Second, insofar as we may specify sides in the quarrel between Ancients and Moderns, the scorecard of disciplinary affiliations does not identify the players of this particular game. In particular, the ranks of Modernists did not include only the new scientific scholars, but also encompassed many prominent intellectuals from literary and other humanistic callings, including the theologian Hakewill.

As a closing example of interdisciplinary medley among the Moderns, and to forsake anglophonic parochialism by a short jog across the Channel (for the so-called quarrel of Ancients and Moderns broke at the same time, and with equal intensity, in both England and France), the story of a remarkable French family should forestall, within its own microcosm, any temptation to view this important historical episode as a dichotomous battle between science and the humanities. If the later revolutionary motto of *liberté, égalité, fraternité* could ever be aptly applied to any minimal group of three, then I nominate the

Perrault brothers as exemplars of all three virtues—the last by a literal bond of biology beyond their choice, but the first two by their own splendid accomplishments. A fourth brother became a noted theologian, a supporter who remains on the sidelines of this particular account.

Claude Perrault (1613–1688), the most prominent scientist among the brothers, joined the extensive ranks of martyrs to his profession—a tradition admirably initiated by the most prominent of Ancients, when Pliny died in the eruption of Mount Vesuvius in A.D. 79. For Perrault expired, at age seventy-five, albeit in a peculiar way that scarcely evokes a conventionally heroic image of death in battle—from an illness contracted after dissecting a camel. Among Claude's numerous talents, he served on the committee that redesigned the eastern façade of the Louvre under Louis XIV. But his chief fame, arising from his medical training, resides in a grand zoological project that he conceived and directed for many years: the establishment of a committee of experts within the Royal Academy of Sciences of Paris, convened to dissect and describe the major forms of vertebrate life by objective procedures of unparalleled care and rigor—particularly by doing each dissection in the presence of several skilled biologists who could reach consensus about their results, and by observing, when available, several specimens and not assuming that a single individual must represent all general features of its type (see figures 17 and 18 for the frontispiece of their book and an example of their classically inspired mode of illustration).

Their resulting volume, published anonymously to emphasize the collective and objective nature of the program, bears the triumphant and lengthy title in my English translation of 1702: *The Natural History of Animals, Containing the Anatomical Description of Several Creatures Dissected by the Royal Academy of Sciences of Paris, Wherein the Construction, Fabric, and Genuine Use of the Parts Are Exactly and Finely Delineated in Copper Plates, and the Whole Enriched with Many Curious Physical and No Less Useful Anatomical Remarks, Being One of the Most Considerable Productions of That Academy.*

In his preface, Perrault extols the virtues of repeatable observations, objectively verified by several experts, in the codification of an optimal methodology for the growing Scientific Revolution:

> That which is most considerable in our memoirs is that unblemishable evidence of a certain and acknowledged verity. For they are not the work of one private person, who may suffer him-

Figure 17.

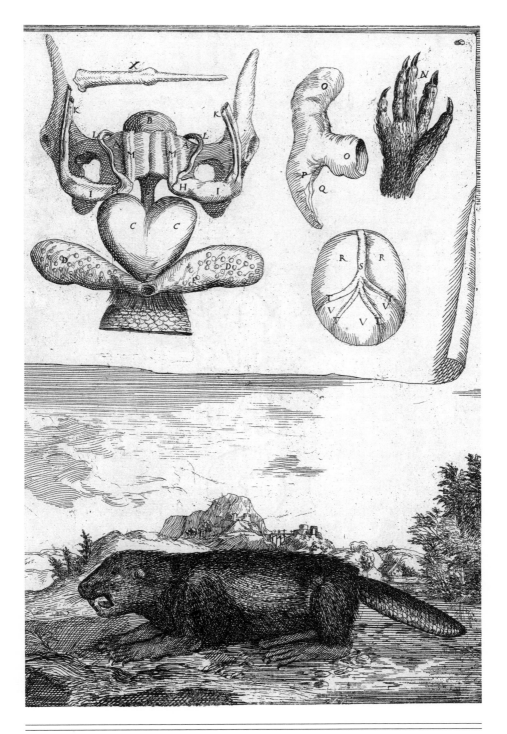

Figure 18.

self to be prevailed upon by his own opinion, who can hardly perceive what contradicts his first conceptions, for which he has all the blindness and fondness which everyone has for his own children. . . . Our memoirs contain only matters of fact that have been verified by a whole Society, composed of men which have eyes to see these sorts of things. . . . Even as they have hands to seek them with more dexterity and success.

Perrault then explicitly weighs in on the Modern side by professing his admiration for the Ancients, while asserting the inevitability of their errors, the progress of modern scientific knowledge, and the right of Moderns to honor their Ancient forebears most truly by correcting their mistakes and thereby advancing a collective enterprise of the ages:

We pretend only to answer some matters of fact, which we advance, and that these facts are the sole powers whereby we would prevail against the Authority of the great persons which have writ before us; seeing that speaking of them with all the respect they deserve, we do own that the defects which are seen in their works are there only because it is impossible to find any thing which has acquired the utmost perfection. . . . For we do think we render a great honor to the merit of the Ancients by demonstrating that we have discovered some small slight errors in their works, than if after that manner of those who distrust their own understanding, and never ground the judgment which they do make of the value of any thing but on prejudices, we should esteem them only because we thought they were done by great Personages, and not by reason of the Knowledge which we have of what they have done well or ill.

As a curious footnote to brother Claude's biology, and as evidence of the hold that the mystique of the Ancients continued to exert over the most committed of Moderns, let me cite the least modest statement that I have ever read in the literature of science. In the closing paragraph of his preface, Claude Perrault, following custom and near necessity, praises the great monarch of France, the Sun King, Louis XIV, who had put his money by his mouth in supporting the committee's labors on vertebrate dissection. Perrault exalts Louis by likening him to Alexander the Great. But why compare an aged monarch in a stable nation with a peripatetic warrior who conquered half of the world and died so young in his boots? The reason for such an odd choice becomes obvious after a moment of reflection. Who had served Alexander as a private tutor in his youth? None other than Aristotle himself. So, as the textual evidence suggests, brother Claude probably selected Alexander not primarily for Louis's sake, but in order to make the comparison of his own scientific work with the labor of Aristotle himself!

> Our memoirs thus being composed, it is to be hoped that they will afford matter for Natural History, which will not be unworthy of the greatest king that ever has been; and that if in this to equal Alexander, as he equals and surpasses him in all other things, he wants [in the old meaning of "lacks"] so great a person as Aristotle, the care which His Majesty has taken to supply this defect by the number of persons which he has chosen for this employ [that is, for dissecting animals and writing this volume of results], and by the order observed to perform the things with an absolute exactness, will make this work, which was undertaken by his command, not inferior perhaps to that which has been done for Alexander.

The second brother, Pierre Perrault (1611–1680), did not work primarily in science, but followed successive careers in law and government service. He did, however, write one great and enduring scientific treatise that established the modern study of hydrology and, in one vital sense, introduced a key proposition of the mechanical worldview as a substitute for an older style of thinking that had symbolized, perhaps more than any other general conviction, the Ancient concept of material reality then under strong attack by the Scientific Revolution. In his 1674 volume, *De l'origine des fontaines* (On the Origin of Springs), brother Pierre advocated the Modernist view of mechanical causality against the Renaissance conviction, dating back to classical times and supported by religious authority as well, that the earth, as the macrocosm and central body of the universe, may be compared in form and action to the microcosm of the human body. (Such a view, for example, underlay Leonardo's geological and geographic writings about landforms and water— see Gould, *Leonardo's Mountain of Clamp and the Diet of Worms*, 1998.)

In this venerable comparison, the bones, blood, breath, and internal heat of the human body—representing the four Greek elements of earth, water, air, and fire—found their counterparts in the rocks, streams, atmosphere, and volcanic heat of our planet. Moreover, just as these elements cycle through the human body, thus maintaining a living entity in steady state, so too must their earthly counterparts cycle through the planet, also (and therefore) construed as an organic and self-sustaining object. Under this concept, the water flowing in streams from the mountains to the seas must then ascend through underground channels (or some other internal system) to the tops of mountains, thus to repeat their descent and maintain the cycle. The "obvious" alternative that we all understand and recognize as factual today—that water returns from the seas to the mountains by evaporation and rain—could not suffice, or even be conceived, under the controlling analogy of microcosm and macrocosm, for the blood of the human body flows through internal channels, and the water of the earth must behave in a similar manner.

Leonardo and others knew about evaporation and rain, but they regarded this source of water as trivial and entirely insufficient to resupply the mountain streams (which must therefore be fed by internal pumping from channels analogous with human blood vessels). Brother Pierre won his small but lasting place in the history of science by proving, and providing numbers and measurements for the Seine to back up his claim, that rainwater could indeed supply all rivers, and that no recourse to internal channels need be sought. The known, and measurable, mechanical forces of evaporation and precipitation thus prevailed over an organic analogy that had nourished centuries of human belief (but not the earth).

Nonetheless, I rest my case for ecumenicism among supporters of the Modern side in this grand debate, and particularly for the allegiance of scientists and humanists—all the varied skills of foxes, united by deepest ties, blood itself in this case, to secure the hedgehog's one great goal of a good and examined life—upon the third and most famous brother, Charles Perrault (1628–1703), the leading literary light of this extraordinary family. Brother Charles became a major figure in the Académie Française, and one of the most celebrated literary men of his time. We may remember him best today—and why not?—as the author, in 1697, of a collection of stories for children entitled *Contes de ma mère l'oye,* or Tales of Mother Goose. But, in his own time, brother Charles won more renown for his vigorous defense of the Moderns, carried out largely among the literati and within the august Académie, in France's intense version of the debate between Ancients and Moderns.

As the *Encyclopaedia Britannica* succinctly states: "In 1671, he was elected to the Académie Française, which soon was sharply divided by the so-called quarrel between the ancients and moderns. Perrault supported the modern view that as civilization progresses, literature evolves with it, and that therefore ancient literature is inevitably more coarse and barbarous than modern literature." In his 1687 poem *Le Siècle de Louis le Grand* (The Age of Louis the Great), brother Charles explicitly praised his colleague Molière as an example of literary grace and perfection that had not, and could not have, been achieved by the writers of Antiquity.

And thus, from life breathed into dead camels (and attained by mortal sacrifice in reverse), to the quenching flow of rainwater, to the grand awakening of a sleeping beauty in our mental aptitude for novelty, these three brothers, spanning the full range of science and the humanities, spoke in equal fraternity for the Modern liberty to move on, and not always to look back.

DICHOTOMOUS PERILS IN FOUR SEQUENTIAL STAGES

From the dawn of recorded human rumination, our best philosophers have noted, and usually lamented, our strong tendency to frame any complex issue as a battle between two opposing camps. Around A.D. 200, for example, Diogenes Laertius cited the dictum of his illustrious forebear Protagoras from the fifth century B.C., a statement already boasting a pedigree of some seven hundred years: "There are two sides to every question, each exactly opposite to the other." Our standard epitomes for the history and social impact of science—with the relationship between science and the humanities as the particular instance highlighted in this book—have consistently followed this preferred mental scheme of dichotomization, although the chosen names and stated aims of the battling armies have changed with the capricious winds of fashion and the evolving norms of scholarship. In Part I, I listed several sequential versions of this supposed dichotomy between the goals of science and the opposing beliefs and practices of humane learning and social convention. In this part I return to these four putative skirmishes in a phantom war—as I try further to expose and understand this false, destructive, and deeply entrenched habit of ordering our categories as oppositional pairs (rather than seeking the virtues of *e pluribus unum* by hybridizing the fox and the hedgehog).

I regard this apparently ineluctable human propensity to dichotomize—the only reason, in my view, that we ever developed a model of opposition between science and the humanities in the first place—as too pervasive and powerful to represent a mere social convention, favored at certain times and by certain types of cultures. I also doubt that anyone would ascribe our predilection for dichotomy to nature's objective factuality—as if our strategy of naming equal and opposite halves expresses an inherently "correct" principle of order for subdividing most classes of objective natural phenomena. I do not deny, of course, that some admittedly basic aspects of our lives suggest a natural parsing into two contrasting clumps, despite some well-recognized fuzziness at the boundaries—with night and day, and male and female, as the primal dichotomies of our external and internal order respectively. (Edmund Burke, the great British statesman, and supporter of America's Revolution despite his generally conservative view of life, remarked sardonically that although no one can draw a sharp line of division—for dawn and twilight designate short zones of intermediary—light and darkness are, on the whole, tolerably distinguishable.)

But, as we consider the totality of similarly broad and fundamental aspects of life, we cannot defend division by two as a natural principle of objective order. Indeed, the "stuff" of the universe often strikes our senses as complex and shaded continua, admittedly with faster and slower moments, and bigger and smaller steps, along the way. Nature does not dictate dualities, trinities, quarterings, or any "objective" basis for human taxonomies; most of our chosen schemes, and our designated numbers of categories, record human choices from a cornucopia of possibilities offered by natural variation from place to place, and permitted by the flexibility of our mental capacities. How many seasons (if we wish to divide by seasons at all) does a year contain? How many stages shall we recognize in a human life?

I strongly suspect that our propensity for dichotomy lies deeply within our basic mental architecture as an evolved property of the human brain—and not as a particularly adaptive trait, either, at least at this point in our history. Claude Levi-Strauss and his school of French structuralism have developed their theory of human nature and social history under the premise that we have evolved an innate propensity for dichotomous classification as our basic cognitive tool for ordering the complexities of both nature and culture. We may begin with empirically defendable divisions of male versus female and night versus day. But we then extend these concrete examples into greater and more subjective generalities of nature versus culture ("the raw and the cooked" of Levi-Strauss's famous book), or spirit versus matter (of philosophical dualism), or the beautiful versus the sublime (in Burke's theory of aesthetics); and thence, and now tragically, into ethical valuation, anathematization, and, sometimes, warfare and mass destruction. For when we add the weight of conscious judgment—another uniquely (and often dangerously) evolved peculiarity of our species—to a simple division by appearance, we turn a formal dichotomy into a moral distinction of good and bad, a transition that can easily slip further into political tragedy, or even genocide, as good and bad intensify into the godly who must prevail versus the diabolical, ripe for burning.

One can speculate about the putative evolutionary basis of such a strong tendency for dichotomization. I rather suspect that this innate propensity represents little more than "baggage" from an evolutionary past of much simpler brains built only to reach those quick decisions—fight or flight, sleep or wake, mate or wait—that make all the difference in a Darwinian world of nonconscious animals. Perhaps we have never been able to transcend the mechanics of a device built to generate simple twofold divisions, and have had to construct our greater complexities upon such a biased and inadequate mental substrate.

I freely confess my negative, and somewhat cynical, feelings about the fallacies (and sometimes even the viciousness) of dichotomization as our usual framework for characterizing the never-ending struggles of academic life—often so silly in their pretentious and vainglorious rancor, especially when honest moments force our admission that degree of public recognition, and differential access to parking spaces, rather than serious issues of intellectual content, usually underlie the intensity of expressed feelings. Viewing the question in its historical amplitude, the most persuasive argument against a concept of "natural" and inherent conflict between science and the humanities may well rest upon the peculiar circumstance that not a single episode in the four successive rounds of this supposed struggle provides any decent evidence for genuine dichotomous opposition, but rather illustrates the far greater complexity, artificiality, contingency, and shifting allegiances of our taxonomies for academic disciplines. So if "science" and the "humanities" cannot be construed as sufficiently stable entities locked in tolerably continuous struggle over genuine and persisting differences of intellectual note, then I suspect that our strong impression of lasting conflict only records our simplistic imposition of phony dichotomous models upon a much different, and far more subtle, story of substantial and fruitful interaction amid instances (or even periods) of misunderstanding and occasional strife.

1. *Ancients and Moderns in the seventeenth and eighteenth centuries.* I have already discussed at some length how many early leaders of the Scientific Revolution boosted the Modern cause by asserting the power of new knowledge, won by observation and experiment, over the Renaissance penchant for recovering Ancient wisdom as a best recipe for intellectual growth—an argument especially well embodied in Bacon's paradox and Newton's aphorism. But the celebrated battle of Ancients and Moderns cannot be read as a dichotomous struggle with an alternative and fully adequate mapping as scientists (Moderns) versus humanists (Ancients)—that is, as an initial skirmish in a more extensive and continuing conflict of science versus the humanities. This simplistic double dichotomy fails by any legitimate criterion, as noted at several earlier points of this text. First of all, many of the greatest naturalists of Western history, particularly during the Renaissance heyday of the fifteenth and sixteenth centuries, followed the Ancient line, with emphasis upon linking modern knowledge of organisms to Aristotle's and Pliny's evidently superior but incompletely preserved insights. Gesner and Aldrovandi, who became the "whipping boys" of seventeenth-century empiricists in the Scientific Revolution (see comments of Grew and Ray on pages 39–47), occupy the first rank both as allies of the Ancients and as superb naturalists.

Second, virtually all leaders of the Scientific Revolution, as befitted general notions of a decent education in those days, learned the standard corpus of Latin and Greek writings, and revered (and liberally quoted) those works, even in their defenses of Modern observational methods. Third, the core of the conventional debate of Ancients and Moderns did not rest upon the argument that new scientific methods could win knowledge heretofore unavailable. Rather, supporters of the Ancients advanced the different and more subtle argument that science's proper insistence upon novel discovery could not be transferred to a literary claim that, by the same token, Modern forms of writing must also surpass Ancient styles because, as a general principle, everything gets better through time. These literary Ancients, in fact, made a proper distinction between the accumulative character of science and the absence of a similar basis for confidence about improvement in the more subjective domain of literary style.

The core of the debate between Ancients and Moderns, after all, resided in a literary struggle, not in a contest between science and the humanities. The *Encyclopaedia Britannica* article notes that Modern literati might have ripped off the successes of science to make an analogistic argument for their humanistic claim, but the basic struggle did not pit science against the humanities:

> The "ancients" maintained that classical literature of Greece and Rome offered the only models for literary excellence; the "moderns" challenged the supremacy of the classical writers. The rise of modern science tempted some French intellectuals to assume that, if Descartes had surpassed ancient science, it might be possible to surpass other ancient arts. The first attacks on the ancients came from Cartesian circles in defense of some heroic poems . . . that were broadly based on Christian rather than classical mythology. . . . Eventually two main issues emerged: whether literature progressed from antiquity to the present as science did [note the acceptance of purely scientific progress by both sides], and whether, if there was progress, it was linear or cyclical.

2. *The warfare of science and religion: a nineteenth-century invention.* The "battle of the books" between Ancients and Moderns, falsely interpreted as an attempt to suppress the early development of modern science, has long faded from public memory and overt influence. But a second episode in the phony war of dichotomies between advancing science and suppressing forces of academic or social convention continues to exert a strong and pernicious influence upon popular culture—the late-nineteenth-century proposal that a "warfare" between science and religion set the primary dynamic of historical change in the Western world. (At least I can assert, speaking personally, that folks of my generation learned this model in the public schools of my youth, although my buddies in parochial school probably received a different line.)

The origin of this influential model can be traced, broadly, to a strong anticlerical movement within late-nineteenth-century rationalism and, more specifically, to two of the greatest success stories in nineteenth-century publishing, despite the entirely different aims of the two books (see page 29 for an earlier citation). In 1874 the physician and amateur historian J. W. Draper

published his *History of the Conflict Between Science and Religion.** A genera-
tion later, in 1896, Andrew Dickson White, the first president of Cornell
University, published his magisterial two-volume work, *A History of the
Warfare of Science with Theology in Christendom.*

Draper, following a lamentable tradition in the history of American prejudice, wrote his book as a Protestant "old American," fearful of Catholic influence, as particularly expressed in the foreign and proletarian origins of most
American Catholics. His book, little more than an anti-papist diatribe, argued
that the liberal spirit of Protestantism could make peace with the beneficial,
and in any case ineluctable, advance of science, whereas dogmatic Catholicism
could reach no such accommodation and had to be superseded or crushed.

Draper expressed this thesis of dichotomous opposition in no uncertain
terms:

> Then has it in truth come to this, that Roman Christianity and
> Science are recognized by their respective adherents as being
> absolutely incompatible; they cannot exist together; one must yield
> to the other; mankind must make its choice—it cannot have both.

White, in strongest possible contrast, wrote as a friend of science and an
even greater champion of religion in its proper spirit and domain. In founding Cornell as a nonsectarian university, White had been frustrated by the
opposition of so many local clergy, who could not abide a liberal institution

---

*As an odd footnote to history, Mr. Draper (then so influential, but now largely forgotten)
had previously surfaced within my world of evolutionary biology at a particularly dramatic
moment in 1860. We all know the famous story of T. H. Huxley's confrontation with
Bishop Samuel (aka "Soapy Sam") Wilberforce over Darwin's heresy, published the year
before, in 1859—although we usually tell the tale in apocryphal form as yet another triumph of advancing science in its dichotomous warfare with religion. (Indeed, the conservative Wilberforce had no love for evolution, but many liberal theologians could be counted
among Darwin's strongest supporters.) This confrontation has usually been described as a
planned and formal debate between the antagonists. In fact the exchange occurred as a spon-
taneous altercation (if not entirely unexpected, given the characters involved and their antic-
ipated attendance) during the discussion period following a formal address by the same Mr.
Draper at the annual meeting of the British Association for the Advancement of Science.
Draper spoke on "The Intellectual Development of Europe Considered with Reference to
the Views of Mr. Darwin."

of higher learning in their midst. White, a dedicated and ecumenical theist, therefore wrote his book to persuade his fellow believers that the beneficial and unstoppable advances of science posed no threat to genuine religion, but only to outmoded dogmas and superstition. White stated this thesis in a famous passage:

> In all modern history, interference with science in the supposed interest of religion, no matter how conscientious such interference may have been, has resulted in the direst evils both to religion and to science. . . . On the other hand, all untrammeled scientific investigation, no matter how dangerous to religion some of its stages may have seemed for the time to be, has invariably resulted in the highest good both of religion and of science.

This model of warfare between science and religion—surely the most powerful analog of the conflict between science and the humanities as a false dichotomy for the history of Western knowledge—fails on both possible rationales: as a defendable antithesis in logic, and as an accurate description in history. I have made the general argument in my book *Rocks of Ages* (Ballantine, 1999), a book that expresses the consensus of a great majority of professional scientists and theologians, not an original formulation from my pen. In briefest summary, no dichotomous opposition can exist in logic because science and religion treat such different (and equally important) aspects of human life—the principle that I have called NOMA as an acronym for the "non-overlapping magisteria," or teaching authorities, of science and religion. Science tries to record and explain the factual character of the natural world, whereas religion struggles with spiritual and ethical questions about the meaning and proper conduct of our lives. The facts of nature simply cannot dictate correct moral behavior or spiritual meaning.

The warfare of science and religion fails equally badly as a description of history. First of all, no one could possibly defend such a model for the founders of the Scientific Revolution in the seventeenth century, as the sincere religious convictions of these men can scarcely be doubted (and genuine atheism enjoyed no popularity at all among scholars of the time). At most, one might entertain a suspicion about Descartes's private attitudes, as his invocations of God do seem minimal and a bit *pro forma* (although not necessarily insincere on this account). But I cannot think of another leading seventeenth-century scientist whose life or works convey the slightest doubt about the strength and importance of their theistic beliefs.

As so many scholars have documented, the standard episodes in the supposed warfare of science and religion are either greatly distorted or entirely fictional. For example, the historian J. B. Russell (*Inventing the Flat Earth*, Praeger, 1991) devotes an entire book to showing how Draper, White, and other architects of the "warfare" model simply invented the old tale of Columbus's brave conflict, as a scientifically savvy navigator, against religious authorities who insisted that he would sail off the edge of a flat earth. In fact, Christian consensus had never lost or challenged Greek and Roman knowledge of the earth's spherical shape. Columbus did hold a celebrated dispute with clerics at Salamanca and other places, but no one questioned the earth's roundness. (His interrogators wore clerical garb because most Spanish scholars at that time were trained, ordained, and employed by the Church, and his adversaries included the best astronomers and geographers of his time and place.) Moreover, his questioners were right, and Columbus entirely wrong. The debaters argued about the earth's diameter, not its shape. Columbus, as his clerical critics correctly documented, had greatly underestimated the size of the earth and could never have reached the Indies by sailing west. Columbus won his lucky and lasting fame only because a large and previously unknown landmass lay in a convenient halfway position. (Native Americans received the epithet of "Indians" as a consequence of Columbus's error.)

Even the canonical tale of Galileo's forced recantation in 1633 cannot stand as an episode in a war between science and faith. Urban VIII remains a villain, and Galileo a hero, in my book, but Galileo was also a frightfully undiplomatic hothead who brought unnecessary trouble upon his own head. He had, after all, received an official imprimatur for publishing his book on Ptolemy versus Copernicus. Church authorities only required that he present an "honest" debate between the two sides, and that he depict heliocentrism as a mathematical hypothesis rather than as empirical truth—a "polite fiction" that would still have won the day for Copernicus. If Galileo had so proceeded, the Copernican view would have triumphed by the inherent character of its superior arguments. Instead, Galileo couldn't resist his urge to ridicule the Ptolemaic opposition by awarding the defense of this position to a character named Simplicio, and by providing him with arguments that matched his name in acumen. No monolithic "church" condemned Galileo, and the considerable cadre of ecclesiastical scientists mostly deplored, if necessarily in silence, the fate of a dear colleague who, as they well knew, had spoken truly and with no antireligious intent. (See *Galileo Courtier* by Mario Biagioli, University of Chicago Press, 1993, for a subtler view of the Galileo affair.)

The late-nineteenth-century formulation of the warfare model arose directly from surrounding contingencies of the time—including the deeper challenge of Darwinian theory to traditional views about the origins of our species, and the occupation of the papacy by the increasingly bitter and deeply conservative Pio Nono (Pope Pius IX, who wins no stars in my book of heroes, but whom I regard as one of the most fascinating figures of the nineteenth century)—and not from any greater validity gained by the dichotomous model in the light of Darwinian challenges. And so the debunking continues right to our present moment, when the most celebrated and supposed example of warfare between science and religion in our times—the attempt by biblical literalists to ban or dilute the teaching of evolution in America's public schools—cannot be so characterized in any fair or accurate account. The majority of professional theologians, including numerous explicit statements throughout the past fifty years of papal pronouncement, from the conservative Pius XII to John Paul II, support the factuality of evolution, and recognize that no aspect of empirical nature can challenge the legitimate role of religion in ethical and spiritual domains outside the logic and authority of science. Rather, the public fight against evolution has been carried out by a small, if vocal and locally powerful, minority of fundamentalists who proclaim the literal truth of the Bible—not a popular idea, to say the least, among most religious people these days. The group that successfully joined as plaintiffs to challenge the Arkansas creationism law in the early 1980s *(McLean v. Arkansas),* thus beginning a series of legal moves that culminated in a Supreme Court victory in 1987, included more theologians than scientists.

3. *Two cultures in the Cold War years.* In 1959, when I was an undergraduate at Antioch College and still wallowing in a naively youthful assumption that scholarly debates packed more excitement, and certainly more potential enlightenment, than any other form of struggle with the possible exception of the World Series (then a sore point, as two of New York's three teams had just departed for greener prospects—and I don't mean chlorophyll—in California), C. P. Snow initiated the mother of all academic shouting matches by presenting his utterly inoffensive and, in retrospect, rather dull Rede Lecture at Cambridge, titled "The Two Cultures." The original version received its share of press, but I doubt that this episode in constructing dichotomy between science and the

humanities would ever have become such a *cause célèbre* if Britain's most famous and most acerbic literary critic, F. R. Leavis, had not launched, in 1962, the most intemperate counterattack in the history of modern squabbling. (Obviously, in some irreducibly visceral sense, no one can face such a barrage of name-calling and deprecation with equanimity, but a little cool reflection on the virtues of both attendant publicity and overwhelming sympathy should quickly dispel any gloom. How could C. P. not benefit, and with a chuckle, in remembering the famous words of Isaiah 1:18: "Come now, and let us reason together, saith the Lord: Though your sins be as scarlet, they shall be as white as snow.")

Snow did not fare so well in the most effective critique published later in 1962 by the American literary scholar Lionel Trilling, who inflicted many of Leavis's strong bites minus the *ad hominem* barks that had won so much sympathy for Snow. Remembering my own enthusiasm and close following of this debate during my undergraduate years (I left Antioch for further study at Columbia in 1963), my rereading of it, as I prepared to write this book, left me with a feeling of disappointment and much ado about nothing.

In arguing that academic life had become riven by a split of scholars into camps of suspicion, disrespect, and mutual incomprehension, and in designating the sides of his putative dichotomy as "literary intellectuals" versus "scientists" (with physical scientists the "most representative"), I believe that Snow had identified a local English phenomenon—and largely a snooty Oxbridge parochialism at that—and elevated his observations into a fallacious general case. Snow had begun his career in science, and ended both in university administration and as a respected novelist for a series of books centered on the minidramas of academic life, and collectively titled *Strangers and Brothers*—so he had lived intensely and professionally in both worlds, and surely knew their inner workings. But I can't help thinking that he falsely equated a particular brand of haughty, hidebound, largely upper-class, traditional British literary culture with the much larger and more variegated community of humanists, and that he failed to realize—even while stating the point—that the British system of disciplinary specialization at such an early age accentuated both the parochialism of allegiance and the ignorance of other fields to an extreme level among Western nations. But to grant Snow his own words in asserting his thesis at the outset of "The Two Cultures":

> I believe the intellectual life of the whole western society is
> increasingly being split into two polar groups: . . . At one pole we

have the literary intellectuals, who incidentally while no one was looking took to referring to themselves as "intellectuals" as though there were no others. I remember G. H. Hardy [the great mathematician] once remarking to me in mild puzzlement, some time in the 1930s: "Have you noticed how the word 'intellectual' is used nowadays? There seems to be a new definition which certainly doesn't include Rutherford or Eddington or Dirac [the leading physicist of the day] . . . or me. It does seem rather odd, don't y'know." Literary intellectuals at one pole—at the other scientists, and as the most representative, the physical scientists. Between the two a gulf of mutual incomprehension—sometimes (particularly among the young) hostility and dislike, but most of all lack of understanding. They have a curious distorted image of each other. Their attitudes are so different that, even on the level of emotion, they can't find much common ground.

In my view, Snow's thesis suffers from two fatal flaws, despite its success in promulgating the most influential twentieth-century claim for dichotomous opposition between science and the humanities. First, as discussed above, I believe that Snow falsely extended a local British phenomenon into a claim for global pattern. Second, Snow, by his own later recognition, mixed two quite different and independent points in the central thrust of his argument, and their incoherence seriously compromises the logic of his entire case. With fully good heart and intentions, but with a bit of British paternalism, Snow added a political argument to his basic thesis about science and literature. He recognized the disparity between rich and poor nations as the most unjust and incendiary feature of modern life. His concern became so intensified in his mind that he made one of the worst predictions ever printed about our recent millennial transition:

> This disparity between the rich and the poor has been noticed. It has been noticed, most acutely and not unnaturally, by the poor. Just because they have noticed it, it won't last for long. Whatever else in the world we know survives to the year 2000, that won't. Once the trick of getting rich is known, as it now is, the world can't survive half rich and half poor. It's just not on.

The debate on "the two cultures" stemmed largely from this forgotten second section of Snow's thesis, rather than from the first part with its basic claim for dichotomy. In fact, by now in our times, both parts have been pretty much forgotten. Most of my scientific colleagues could identify Snow, and could probably even cite the title of his famous address. But, although the lecture remains in print, I hardly know anyone who has read this short document in recent years. In a sense, Snow's catchy and dichotomous name became too successful, for everyone remembered the title, and the one-sentence caricature, while forgetting the argument and then ignoring the text itself.

Intensity of debate on this second half arose from the legitimate feeling of humanists that Snow, despite his undeniably good intentions, had inexcusably simplified the problem of poverty in the developing world, and had added insult by touting his own scientific colleagues as quick and sole saviors. For Snow did argue that the end of poverty would be achieved by little more than adequate training of enough local scientists and engineers—a simple technological fix, easily achievable in a few years. He wrote of China with a healthy denial of racism, but with a simplistic disregard of cultural and political issues:

> For the task of totally industrializing a major country, as in China today, it only takes will to train enough scientists and engineers and technicians. Will, and quite a small number of years. There is no evidence that any country or race is better than any other in scientific teachability: there is a good deal of evidence that all are much alike. Tradition and technical background seem to count for surprisingly little.

Snow acknowledges that this expertise must be imported from the West, and he does add a small warning about paternalism as his only slight caveat about social difficulties. But he then falls immediately back into naive optimism, accompanied by yet another zinger (as read by his humanistic colleagues) about the inherent ability of scientists, as opposed to other folks, to work in this cooperative and sensitive manner with others:

> Plenty of Europeans, from St. Francis Xavier to Schweitzer, have devoted their lives to Asians and Africans, nobly but paternally. These are not the Europeans whom Asians and Africans are going to welcome now. They want men who will muck in as col-

leagues, who will pass on what they know, do an honest technical job, and get out. Fortunately, this is an attitude which comes easily to scientists. They are freer than most people from racial feeling; their own culture is in its human relations a democratic one. In their own internal climate, the breeze of the equality of man hits you in the face. That is why scientists would do us good all over Asia and Africa.

In 1963, largely in response to the firestorm initiated by Leavis and Trilling, Snow published a reassessment and update of his defining claim for dichotomy of science and the humanities in our times—*The Two Cultures: A Second Look.* His almost courtly, sometimes sardonic, always firm yet utterly unpetty commentary on the criticism surrounding his initial essay won nothing but plaudits for style and fairness. I particularly appreciated his wry summary: "From the beginning, the phrase 'the two cultures' evoked some protests. The word 'culture' or 'cultures' has been objected to; so, with much more substance, has the number two. (No one, I think, has yet complained about the definite article.)"

But much of Snow's rumination then moves from defense to acknowledgment and self-criticism. In particular—and providing my major reason for treating Snow's work at length in this critique of dichotomy, and in brief for hybridizing the fox and hedgehog—Snow effectively surrenders, and reverses his position on what had been, after all, the motivating assumption of his original argument: the validity of a dichotomous parsing of intellectual life into contrarian literary and scientific camps (however much Snow deplored the opposition and hoped to facilitate its easing or disappearance). Even in the original essay, Snow had sensed, and acknowledged, the problems of our all-too-convenient divisions by two:

> The number 2 is a very dangerous number: that is why dialectic is a dangerous process. Attempts to divide anything into two ought to be regarded with much suspicion. I have thought a long time about going in for further refinements: but in the end I have decided against. I was searching for something a little more than a dashing metaphor, a good deal less than a cultural map: and for those purposes the two cultures is about right, and subtilising any more would bring more disadvantages than it's worth.

But, by 1963, Snow had reassessed this basic decision and its resulting model. He had apparently recognized just how severely he had caricatured his two sides by choosing extremes as their exemplars in each case—upper-class, literary Oxbridge dons to stand in for all the humanities, and devotees of the "hardest" quantitative and experimental methodologies in physical science to represent the full range of folks who study factual nature in all its ways and manifestations. In the intervening years, Snow had obviously explored the enormous middle ground between these artificial end points—not just a few oddballs in a small transitional zone, but a vast mass of scholars, probably constituting the great majority in a continuum that certainly cannot be depicted as a dichotomy defined by the rare extremes at each terminus.

Moreover, I think Snow now realized that although the continuum of a single axis suggested a richer and truer model than a dichotomy, intellectual life spread out in too many directions to depict along a single axis in any case. I regard this admission as an honorable surrender, a throwing of the towel into this particular academic boxing ring. Snow's expansion suggested that what we roughly characterize as the "social sciences" should probably be formulated as a third culture, thus implying a fourth, a fifth, and, by extension, the death of the dichotomous model that had sparked all the controversy in the first place! Thus, I view the history of discussion about Snow's "Two Cultures" as a lesson in the fallacies and dangers of dichotomy (while I obviously do not deny the value of such simplification in provoking discussion and better resolution). Snow wrote:

> I have been increasingly impressed by a body of intellectual opinion, forming itself, without organisation, without any kind of lead or conscious direction, under the surface of this debate. This body of opinion seems to come from intellectual persons in a variety of fields—social history, sociology, demography, political science, economics, government (in the American academic sense), psychology, medicine, and social arts such as architecture. It seems a mixed bag: but there is an inner consistency. All of them are concerned with how human beings are living or have lived—and concerned, not in terms of legend, but of fact. I am not implying that they agree with each other, but in their approach to cardinal problems—such as human effects of the scientific revolution, which is the fighting point of this whole affair—they display, at the least a family resemblance.

I ought, I see now, to have expected this. I haven't much excuse for not doing so. I have been in close intellectual contact with social historians most of my life: they have influenced me a good deal: their recent researches were the basis for a good many of my statements. But nevertheless I was slow to observe the development of what, in the terms of our formulae, is becoming something like a third culture. I might have been quicker if I had not been the prisoner of my English upbringing, conditioned to be suspicious of any but the established intellectual disciplines, unreservedly at home only with the "hard" subjects. For this I am sorry.

4. *Postmodernism and the millennial "science wars."* As the debate over C. P. Snow's version of the dichotomous opposition between science and humanities died down and passed into the academic limbo of lost fashion (that is, still available for historical chronicles but not for current passions), an even more general and, if anything, apparently more contentious episode emerged in what American vernacular calls the "same old same old."

As my own cynicism grew after my undergraduate fascination with the apparent depth of Snow's "two cultures" debate (on more sanguine days, I would substitute "wisdom" for "cynicism" as the descriptor of my maturation), I came to realize that most of the starkness and uncompromising opposition in all these episodes of dueling dichotomy arises not from any position actually taken by either party in the debate, but rather from the strawmen of extremity invented by one side to discredit the other and win the argument by ridicule. As an old, and factual rather than cynical, principle of human affairs, the victors get to write history, and their willful mischaracterizations, invented in the heat of battle, tend to persist even when, in another (and far better) convention of human affairs, generosity should prevail as a correlate of victory.

It took me many years, and many incidents of puzzlement, to recognize this particular trick and trope. I would imbibe the myth of victors about the position just vanquished. I would then search the documents of the vanquished for affirmation—and never find any claim even close to the exaggerated version that made the victory of my side so sweet and so necessary. Instead of suspecting the victors of caricature, I just looked harder for the position that my side had imputed to their enemies. Only much later did I grow the guts, and acquire the intellectual maturity, to suspect, and then virtually to prove by more assiduous study, that victors often distort their opponents' views into absurd extremes.

If I may enter the confession booth and admit an embarrassing example from my very first publication, I wrote an essay on uniformitarianism in geology, making some novel points that I continue to recall with pride. But I had learned from day one of my first course in geology that the "bad guys" of an early-nineteenth-century dichotomy, known (boo, hiss) as "catastrophists," were antiscientific theological apologists who argued for paroxysmal geological change at global scale because they dogmatically accepted both the efficacy of miracles and the six-thousand-year literal chronology of Genesis. But I read and read, and never found a hint of affirmation for either claim. Rather, all leading catastrophists seemed to agree with the uniformitarians about an ancient earth. They also shunned miracles as outside the course of natural law, and therefore incapable of scientific explanation. In fact the catastrophists seemed to be making the theoretically honorable (if factually dubious) point that geological dynamics on our ancient earth had been primarily paroxysmal but entirely natural—rather than gradual and accumulative as the uniformitarians favored.

But as a young undergraduate publishing his first paper (Gould, 1965), I simply lacked the courage to believe my own discovery. So I kept reading until I found one quotation by a catastrophist that could be read as a theological apology. I cited this single statement and assumed that I must have missed all the others. After all, how could such a venerable and monolithic account from the standard literature be so mistaken? Now I know better; but I wish I had possessed the courage to say so in 1965. Still, the moving finger writes, and having writ, moves on. . . .

I have been led to recall this embarrassment of my fledgling days in science for a largely emotional reason. I watched the development of this fourth and latest episode in dichotomous battle between science and the humanities from the vantage point of a professional life as both a practicing scientist and a literate commentator and general essayist on the history and impact of science—that is, as a decently fledged adult with a reasonable range of foxy skills and a strong brief for the hedgehog's cause, and not, as for the third episode of Snow's "Two Cultures," as an inarticulate beginner. And I watched in sheer frustration (and too much silence, for I should have spoken out far more than I did) as the two perceived sides formed their supposed battle lines in a struggle that soon received an almost "official" designation in pure martial metaphor, as "the science wars." And yet I had never witnessed a clearer example of the "emperor's new clothes" fallacy, for no garments of veracity covered this particular invention. I could only recall a sardonic motto from my undergraduate days in the antinuclear movement: "What if they gave a war, and nobody came?"

In its semiofficial incarnation as a fourth battle in an ancient dichotomy, the "science wars" supposedly pitted a group of radical, self-styled "postmodern" scholars in the humanities and social-science departments of American universities (particularly representing a new field called "science studies") against researchers in the conventional science departments of the same institutions. The postmodern critics—dubbed "relativists"—had supposedly branded science as just one choice among our infinite and inherently subjective ways of human knowing, with no genuine claim upon methods that could validate nature's factuality. Rather, and cynically (if perhaps naively and unconsciously in some cases), scientists invoked this rhetoric of a privileged path to objective knowledge, or so the postmodernists claimed, in order to win funding, power, and influence by bluff. The professional scientists on the other side—dubbed "realists"—denied the validity of any social analysis of scientific practice, and would not even admit that unconscious political and psychological preferences might influence scientific belief (except as clear and correctable failures of individual researchers who had not fully appreciated or applied the proper "scientific method" to their own work). Moreover, these realists supposedly maintained that science alone held the methodological key to any form of knowable truth, and that science, at least in its technological manifestations, lay behind all advance and improvement in the dynamics of Western history.

And so the impression went abroad that scientists themselves, and analysts of science within the humanities, philosophy, and social studies, had locked themselves into an overt struggle—a true "science war"—over any privileged domain of expertise for science, indeed over the very concept of factual truth and scientific progress at all. Extremists on each side presented absurd caricatures of the supposed opposition and, since everyone loves a fight, journalistic commentaries in America's very few outlets for passably intellectual writing described the "science wars" with such verve that an unsuspecting reader might actually have imagined campuses filled with barricades occupied by professors hurling verbal stink bombs.

And yes, if one searched the literature, one could find a few commentaries either unwisely exaggerating the supposed conflict, or easily misreadable as so doing—and so the impression of dichotomy only accelerated, at least for a while. For example, from the literary side, consider this sequence of three statements on the development of "science wars" from Stefan Collini's 1998 introduction to a reprint of Snow's original "Two Cultures" piece. He begins by accurately describing the quite reasonable critiques of historians and sociologists of science:

> A broader programme of the social history of science has concentrated attention upon the role of "external" factors, such as the class origins of scientists themselves, the political and cultural forces steering research in some directions rather than others, and the social and psychological needs catered to by ideals of professionalism and disinterestedness.

Collini then, and still describing rather than judging, continues:

> More radically still, much recent work has been devoted to showing how the very constitution of scientific knowledge itself is dependent upon culturally variable norms and practices; seen in this way, "science" is merely one set of cultural activities among others, as much an expression of a society's orientation to the world as its art or religion, and equally inseparable from fundamental issues of politics and morality.

I don't object to this more radical claim—and scientists should consider this more subtle and pervasive extent of their social role and embeddedness. After all, Collini isn't denying that factual truth exists, and that science might accurately locate some of it. But, particularly in sensitive times, one might excuse an overwrought scientist for drawing such an extended, if unstated, inference—especially when Collini then quotes, on the next page, an even more provocative claim from a leading "relativist," Wolf Lepenies:

> Science must no longer give the impression it represents a faithful reflection of reality. What it is, rather, is a cultural system, and it exhibits to us an alienated interest-determined image of reality specific to a definite time and place.

Now, them's potentially fighting words. Maybe Lepenies (as I actually suspect) only means, in a colorful way, to expose the intrinsic embedding of science, and scientific practice, into the changing norms of surrounding culture—and to expose our willingness to track those norms (whether consciously or not) in attempts to gain political support, as folks do in all fields. Perhaps Lepenies is not denying that, despite these shifting social realities, science can still establish accurate, or at least technologically useful, accounts of the factual world. But one can hardly blame scientists for thinking that Lepenies might be denying any intelligible meaning to the concept of scientific truth at all—hence the epithet "relativist" for this supposed side in a putative fourth episode of dichotomy.

Counterattacks by scientists have been infrequent (see below for the somewhat surprising and largely unappreciated reason), but interesting and sometimes disconcerting. Some of my colleagues have become legitimately disturbed by a few truly silly and extreme statements from the "relativist" camp, largely made by poseurs rather than genuine scholars, and have mistaken these infrequent sound bites of pure nonsense for the center of a serious and useful critique. Then, falsely believing that the entire field of "science studies" has launched a crazed attack upon science and the concept of truth itself, they fight back by searching out the rare inane statements of a few irresponsible relativists (every field, after all, must bear the burden of its own fringe) and then presenting a polemic defense of science, ultimately helpful to no one—for no serious enemy exists in the form described, and no one appreciates a shrill diatribe against a caricature of one's more subtle and genuine concerns (see a striking example of such windmill-bashing in *Higher Superstition: The Academic Left and Its Quarrels with Science,* published in 1994 by P. R. Gross and N. Levitt).

The most clever of scientific counterattacks filled me with both amusement and disquiet. My friend Alan Sokal, professor of physics at New York University, performed an unusual "experiment" to find out if certain social critics even understood the content of the scientific concepts falling under their supposed scrutiny. So Sokal wrote a delicious, but pretty darned transparent, parody, ostensibly claiming that as a formerly unrepentant realist, he had seen the light of relativism and now accepted the defining argument for social construction, rather than objective factual reality, of scientific conclusions.

I don't think that Sokal ever expected his effort to go so far—but, hey, once you're in, you're in, and it becomes *their* job to find you out. So Sokal sent his manuscript, laden with enough laughs and clues to identify the article as pure parody to anyone with a modicum of scientific knowledge, to *Social Text,* a leading journal in the relativist camp of the science wars (by the usual taxonomy of this episode). The editors' pleasure and ardor at the prospect of such a prominent born-again convert obviously canceled their suspicions and critical faculties—and they published Sokal's article as a serious piece, expressing the triumph of their point of view among the enemy. Needless to say, when Sokal immediately admitted his content and intent, thus eliciting gleeful reports on the front page of the *New York Times* and other leading publications throughout the world, the editors of *Social Text* ate a large murder of crows (far beyond those four and twenty blackbirds baked in a pie).

Fine, Sokal had clearly proved a point—but what point? I confess to very mixed feelings about this incident (and Sokal and I have discussed the issue at length, and without resolution because, frankly, I have never been able to sort out my own complex feelings about the affair). The parody was brilliantly done, and the results as funny as could be—and Sokal does dwell on "my" side. But parody is also a very broad and coarse weapon, and its intentions often backfire in a philistine world. Too many people—and I know that Sokal didn't intend or desire such a result—read the incident as a full and general indictment of all social criticism of science, and of any studies in the history of science that stress social context over pure logic of argument. But I, as a practicing scientist, happen to regard the vast bulk of scholarly work in the social analysis of science as not only important and respectable, but as immensely salutary for scientists who rarely think enough about the historical background and immediate social context of their research, and who would therefore greatly benefit from better understanding of these nonscientific influences upon their beliefs and practices.

So had Sokal exposed the entire field of science studies as a bunch of poseurs and braying ignoramuses? I don't think so. Frankly, I think he only exposed the hubris or laziness of the particular editors of *Social Text,* who became so beguiled by apparent support from the "other" side that, despite their complete ignorance of the physics discussed in Sokal's paper, they failed to exercise the standard (and, in most technical journals, formally and absolutely required) procedure of sending the paper to an expert in physics for "peer review." Any physicist would have immediately recognized the parody (and any careful lay reader, not anesthetized by pleasure at the apparent content, should have been suspicious for a hundred different reasons). So, does the publication of Sokal's parody condemn an entire field or merely expose the carelessness of a few chagrined and chastened editors? I see no lesson in the incident beyond this second and smaller outcome, with its strictly limited message. And yet, as I said above, parody can be a dangerous weapon—and many observers dismissed the entire field of history and social analysis of science, an important and productive branch of modern scholarship, because a few practitioners, by their own malfeasance, had been embarrassed and exposed.

Finally, and to prove my point about the nonexistence of these supposed "science wars"—thus exposing the fourth episode of dichotomy as not merely distorted, but truly fictional—let me clinch the argument by revealing a trade secret about my fellow scientists that our little minority with literary pretensions in the business would probably prefer to keep hidden. I do love my colleagues dearly, at least most of them. I stand in awe before their dedication and technical skills. But, to be frank and to put the matter bluntly, the vast majority of scientists are a parochial lot. No one could accuse us of pure one-dimensionality; your average scientist likes to read a diverting book on a long flight, watch the latest movie, and root hard for the home team. Many, perhaps even most, of us are even tolerably intellectual. We will visit a museum or attend a concert without undue protest, and often with pleasure; we may even play a musical instrument with reasonable competence. But the vast majority of us will never—and I mean *never*—even dream about reading technical academic literature from other fields, particularly literature that claims to present deep, critical, and insightful analysis of science as an institution, to reveal the psychology of scientists as ordinary folks with ordinary drives, or to depict the history of science as a socially embedded institution. I mean, why read about it, as written by outsiders, when we live it every single day?

I do not defend—indeed I deplore—this "philistinism lite" so prevalent among my colleagues. But, deplore though I may, the existence of this pervasive tendency cannot be denied. Most scientists have never read a technical work in the history or philosophy of science; and most of my colleagues could not identify a single leader in the field—not Thomas Kuhn or Karl Popper from the last generation, and not any lesser light in the supposed "science wars" of our present moment. Thus no "science war" exists for the most obvious and irrefutable of all reasons: the vast majority of scientists have never heard about the supposed altercation and have no interest whatever in considering a claim so utterly incomprehensible to them as the relativist argument for a social construction, rather than a factual basis, of scientific knowledge. Tell most scientists about the "science wars"—and I have tried this experiment at least fifty times—and they will stare back at you in utter disbelief. They have never encountered such a thing, never read anything about it, and don't care to interrupt their work to find out. Oh yes, the occasional savvy scientist who pals around in urban intellectual circles may engage the "wars" and get pissed off—leading to the expressed anger of Gross and Levitt, or the wry amusement of Sokal. But most of my colleagues know nothing at all about the war supposedly being conducted in (or against) their name. And, as an old motto, previously cited, acknowledges, you can't have a war if one side declines to show up.

I particularly deplore this fourth false episode of putative conflict between the sciences and humanities, because the opposing camps were confected of extreme views held by virtually no one on either supposed side—whereas the actual, and more nuanced, opinions of sensible folks in both the "relativist" and "realist" contingents express important insights that could greatly benefit the understanding of practitioners in the other party, if only the two groups would pay attention to each other, recognize the extreme caricatures as harmful fictions, and learn to appreciate the fair and just emphases of each group: (1) the stress that historians and political analysts of science place upon social construction; and (2) the weight that practicing scientists place upon the extraordinary capacity and success of scientific methods in acquiring reliable and technologically useful knowledge about what can only be called (admittedly by inference, but what else could one infer?) the factual structure of material reality.

In other words,* we must reject the widespread belief that a science war now defines the public and scholarly analysis of this institution, with this sup-

---

*The rest of this section, and the portion of the next section on Bacon's idols, have been taken in part from the technical article that I wrote to initiate the Millennium series of *Science* magazine on "Pathways of Discovery" (issue of January 14, 2000). This general name, chosen by the editors for this highly unusual historical series in America's leading journal for professional scientists, does, I confess, underscore the valid concerns of humanists. For this title highlights an assumption about history as a march to true answers (pathways, construed as basically straight), thus bypassing the insights gained by historians of science about social embeddedness and construction. Moreover, the fact that our key journal deigns to feature a historical series only as a millennial "frill" does underscore the peripheral status of the enterprise in the consciousness of most scientists. This article served as the inspiration and outline for the present book. And I justify this double-dipping into my own work thereby.

posed struggle depicted as a harsh conflict pitting realists engaged in the practice of science against relativists pursuing the social analysis of science. Most working scientists may be naive about the history of their discipline and therefore overly susceptible to the lure of objectivist mythology. But I have never met a pure scientific realist who views social context as entirely irrelevant, or only as an enemy to be expunged by the twin lights of universal reason and incontrovertible observation. And surely no working scientist can espouse pure relativism at the other pole of the dichotomy. The public, I suspect, misunderstands the basic reason for such exceptionless denial. In numerous letters and queries, sympathetic and interested nonprofessionals have expressed to me their assumption that scientists cannot be relativists because a professional commitment to such a grand and glorious goal as the explanation of our vast and mysterious universe must presuppose a genuine reality "out there" to discover. In fact, as all working scientists know in their bones, the incoherence of relativism arises from virtually opposite and entirely quotidian motives. Most daily activity in science can only be described as tedious and boring, not to mention expensive and frustrating. Thomas Edison calculated well in devising his famous formula for invention: 1 percent inspiration mixed with 99 percent perspiration. How could scientists ever muster the energy and stamina to clean cages, run gels, calibrate instruments, and replicate experiments, if they did not believe that such exhausting, exacting, mindless, and repetitious activities could reveal truthful information about a real world? If all science arises as pure social construction, one might as well reside in an armchair and think great thoughts.

Similarly, and ignoring some self-promoting and cynical rhetoricians, I have never met a serious social critic or historian of science who espoused anything close to a doctrine of pure relativism. The true, insightful, and fundamental statement that science, as a quintessentially human activity, must reflect a surrounding social context does not imply either that no accessible external reality exists, or that science, as a socially constructed institution, cannot achieve progressively more adequate understanding of nature's facts and mechanisms.

The social and historical analysis of science poses no threat to the institution's core assumption about the existence of an accessible "real world" that we have actually managed to understand with increasing efficacy, thus validating the claim that science, in some meaningful sense, "progresses." Rather, scientists should cherish good historical analysis for two compelling reasons.

First, real, gutsy, flawed, socially embedded history of science is so immeasurably more interesting and accurate than the usual cardboard pap about marches to truth fueled by universal and disembodied weapons of reason and observation ("the scientific method") against antiquated dogmas and social constraints. Second, this more sophisticated social and historical analysis can aid both the institution of science and the work of scientists—the institution, by revealing science as an accessible form of human creativity, not as an arcane enterprise hostile to ordinary thought and feeling, and open only to a trained priesthood; and the individual, by fracturing the objectivist myth that only generates indifference to self-examination, and by encouraging study and scrutiny of the social contexts that channel our thinking and frustrate our potential creativity. In fact, and speaking now to my colleagues in science, I can cite no better example for the benefit of foxy diversity, gained from reading humanistic analyses of the social role and personal psychology of scientists, in abetting our hedgehog's goal of doing "straight" science even better.

HOW NOW DICHOTOMY— AND HOW NOT

If, in general, dichotomy represents such a false mode for parsing either the structure of nature or the forms of human discourse; and if, in particular, we have erred grievously every time in depicting the history of interaction between science and the humanities as a series of episodes in dichotomous struggle, then why does this fallacy of reasoning, like the proverbial bad penny, keep turning up to poison our understanding and sour our relationships? I would end this critical yet hopeful commentary (for the optimistic side of my being compels me to believe that the exposure of a fallacy can lead to its correction, whatever the odds or the entrenchments) by reiterating three major reasons for the hold of dichotomy upon our schemes and perceptions. The third and most important factor also grants me the literary license to end this meandering section in tight and recursive form by returning to the opening discussion of Francis Bacon, the much misunderstood and underappreciated avatar of the Scientific Revolution, but also a wise social and philosophical critic who, so long ago, presented the best refutation of dichotomy, both in the lesson of his life and the content of his argument.

1. *The turf wars of history.* However tight the logic of respectful separation

may be, and however salutary the benefits of such equal and mutually sup-portive regard might prove, a basic foible of human affairs prevents the achievement of such gracious sharing when the history of turf—whether the prize be actual land and resources or just intellectual space—begins with one side as steward of the totality. No one (or at least no institution in full una-nimity) cedes turf voluntarily, however ultimately beneficial the move and strategy. Thus, if basic human inquisitiveness forces us to ask great questions about why the sky is blue and the grass green, and if, *faute de mieux,* this dis-course fell under the rubric of theology before modern science arose to claim proper dominion over factual aspects of such inquiries about the natural world, then some theologians will resist (while others will see farther and strongly approve) the exit of religion from a domain that never properly fell under its competence.

Similarly, if Renaissance humanists once assumed that their techniques of locating and explicating Ancient texts could best resolve all questions about factual nature, then some adherents to this orthodoxy will resist the legitimate claims of a new institution—modern science—for observation and experi-ment as a more effective pathway to the same goal. With goodwill and the passage of time, these inevitable roilings and suspicions should settle down into an honorable peace based on advantages for both sides (a "win-win" sit-uation in the jargon of our times). But we should probably regard the initial (and heated) skirmishes as unavoidable—the basic theme of the first part of this book, on the "rite and rights of an initiating spring" for modern science. And we should confine our task to deploring and correcting the continuation of such a conflict well beyond this early period of legitimacy—as this inevitable opening move can only become destructive once a novel field has secured its birthright, for generosity and mutual support should then prevail.

2. *The hopes of psychology.* Scientists must understand the limits of their calling for a second practical and powerful reason beyond the first argument above, about turf wars. We live in a vale of tears, and bad things often hap-pen to good people. These unpleasant facts about life cannot be avoided. Therefore, and especially, we need to sustain a realm of human goodness, and a calm place of optimism based on value and meaning, amid realities that we yearn to avoid but cannot deny. Yet our hopes and needs run so high that, until the reality of reiterated experience forces us to bite the bullet and bow to the inevitable, we also try to invest factual nature with the sustaining myths of "all things bright and beautiful," or the psalmist's vain hope (37:25) and

massive self-deception: "I have been young, and now am old; yet have I not seen the righteous forsaken, nor his seed begging bread."

Science can only document these realities that all of us would rather deny or mitigate. And because humans have long practiced a lamentable tendency to slay the innocent messenger of bad news, science does need to specify and defend its role as a messenger and not a moralizer, and then to insist that the message, properly read (admittedly against the hopes and traditions of ages), truly contains seeds of resolution and grounds for genuine optimism. That is, science must insist that, whatever the factual state of nature, our yearnings and quest for morality and meaning belong to the different domains of the humanities, the arts, philosophy, and theology—and cannot be adjudicated by the findings of science. Facts may enrich and enlighten our moral questions (about the definition of death, the beginning of life, or the validity of using embryonic stem cells in biological research). But facts cannot dictate the answers to questions about the "oughts" of conduct or the spiritual meanings of our lives. If we keep these distinctions clear, then nature's unpleasant facts, as ascertained by science, pose no threat to humane studies, and may even foster our discourse in morality and art by posing new issues in different ways.

Still, scientists must recognize and understand how legitimate fear often trumps solid logic to cast unfair suspicion upon a messenger, especially when such a long tradition fuels the false dichotomy and resulting enmity. Thus I do acknowledge how much Wordsworth loved nature, and I do not begrudge his fears, though I must criticize his argument, when he wrote so famously, in a beautiful, but tragically flawed, verse:

> Sweet is the lore which nature brings,
> Our meddling intellect
> Distorts the beauteous forms of things.
> We murder to dissect.

I would only say to the poets that science must dissect as one path to understanding, but never to destroy the beauty and joy of wholeness. And I do regret that some of my colleagues have made rash claims for granting science a decisive role in aesthetic and moral judgment. To all our Wordsworths, I would only grant assurance and strongly affirm that my profession can never challenge, and should only admire, your identification and reverence for those "thoughts that do often lie too deep for tears," to cite the final line of the *Ode on Intimations of Immortality,* judged by Emerson (and I agree) as the finest poem ever written in the English language. I would also remind Mr. Wordsworth that the "host of golden daffodils," his embodiment of joy in nature, grew within my realm and under my rules—and that I experience nothing but pleasure and gratitude in learning about his appreciation and inspiration.

3. *The inborn habits of dichotomy.* I have argued throughout this part that, however intensified by particular reasons of history and psychology, the affliction of dichotomy—the basis for our false, yet persistent, model of opposition between science and the humanities—probably lies deep within our neurological wiring as an evolved property of mental functioning, once adaptive in distant ancestors with far more limited brain power, but now inherited as cognitive baggage. This impediment from our evolutionary past engenders great harm in leading us to misunderstand the complexities that now define our lives and dangers—thus overwhelming whatever benefit dichotomy might still provide in simplifying the immediate cognitive decisions that defined the "do or die" of some ancient forebears, but that now rarely impact our current lives in the same way.

In an admittedly ironic paradox of recursion (the requirement that mind must reflect upon mind in order to break the primary impediment), our best chance for exposing and expunging the fallacy of dichotomous opposition between science and the humanities lies in showing that a powerful myth about scientific procedure—the legend that spawned the impression of science as an objective activity, strictly divorced from all the mental quirks and subjectivities underlying creative work in the humanities—founders on a false assumption best exposed by scrutinizing such intrinsic mental biases as our propensity for dichotomy itself. These universal cognitive biases affect the work of scientists as strongly as they impact any other human activity—perhaps with even greater force because scientists have so firmly enclosed themselves within an ideology that denies the efficacy, or even the existence, of such biases. And what influence can be more pervasive or insidious than a strong effect that cannot be perceived because the rules of the game preclude a proper perception of the problem?

This myth of objectivity—the belief that scientists achieve their special status by freeing their minds of constraining social bias and learning to see nature directly under established rules of "the scientific method"—drives a wedge between science and the humanities, because historians, sociologists, and philosophers of science know that such a mental state cannot be achieved (while they do not doubt the ability of science to gain reliable factual knowledge about the natural world, even if this knowledge must be obtained in curiously roundabout ways by flawed human reasoning); whereas scientists mistake these truthful and helpful analyses by colleagues in the humanities as attacks upon the purity of their enterprise, rather than an intended affirmation that all our mental activities, including science, can only be pursued by gutsy human beings, warts and all (and that we often learn more from the warts than from the idealizations).

If scientists would admit the ineluctable human character of their enterprise, and if students of science within the humanities would then acknowledge the power of science to increase the storehouse of genuine knowledge by working with all the flaws of human foibles, then we could break the hold of dichotomy and break bread together. The first, and in many ways still the best, analysis of the inherent mental biases underlying all scientific work resides in the most important treatise written by Francis Bacon himself—a particularly ironic situation because Bacon's name then became associated with the opposite position that has fueled the flames of dichotomy for centuries. For reasons described just below, the "objective" process of simply recording facts, and then drawing logical inferences from these lists of facts alone, became known, in anglophone jargon, as "the Baconian method," thus tying the name of this avatar of the Scientific Revolution to the myth that then drove a wedge between science and other intellectual activities—not Bacon's intention at all, as we shall see.

For example, in a famous statement from his autobiography, Charles Darwin, with uncharacteristic misunderstanding (or misremembering) of his own life and work, described his initial inklings about evolution: "My first notebook was opened July 1837. I worked on true Baconian principles and without any theory collected facts on a wholesale scale." Of course, Darwin did not, and could not, so proceed. From the very beginning he tested, retested, proposed, rejected, and refined a wide and everchanging spate of theoretical assumptions, until he finally developed the theory of natural selection by a complex coordination of mental preferences and factual affirmation. To refute his own naive claim, I need only restate my favorite Darwinian line, cited several times before: "How odd it is that anyone should not see that all observation must be for or against some view if it is to be of any service."

Bacon's dubious, and wholly undeserved, reputation as the apostle of a purely enumerative and accumulative view of factuality as the basis for theoretical understanding in science rests upon the tables for inductive inference that he included in the *Novum Organum,* the first substantive section following the introduction to his projected *Great Instauration.* Bacon, who has never been accused of modesty, had vowed as a young man "to take all knowledge for my province." To break the primary impediment of unquestioned obeisance to ancient authority (the permanence and optimality of classical texts), Bacon vowed to write a *Great Instauration* (or New Beginning) based on principles of reasoning that could increase human knowledge by using the empirical procedures then under development and now called "science."

Aristotle's treatises on reasoning had been gathered together by his followers and named the *Organon* (tool, or instrument). Bacon therefore named his treatise on methods of empirical reasoning the *Novum Organum,* or "new instrument" for the Scientific Revolution. The "Baconian method," as Darwin used and understood the term, followed the tabular procedures of the *Novum Organum* for stating and classifying observations, and for drawing inductive inferences therefrom, based on common properties of the tabulations.

Perhaps Bacon's tables do rely too much on listing and classifying by common properties, and too little on the explicit testing of hypotheses. Perhaps, therefore, this feature of his methodology does buttress the objectivist myth that has so falsely separated science from other forms of human creativity. But when we consider the context of Bacon's own time, particularly his need to emphasize the power of factual novelty in refuting a widespread belief in textual authority as the only path to genuine knowledge, we may understand an emphasis that we would now label as exaggerated or undue (largely as a consequence of science's preeminent success).

Nonetheless, a grand irony haunts the *Novum Organum,* for this work, through its tabular devices, established Bacon's reputation as godfather to the primary myth of science as an "automatic" method of pure observation and reason, divorced from all sloppy and gutsy forms of human mentality, and therefore prey to the dichotomous separations that have so falsely represented the relations of science and the humanities for more than three hundred years of Western history. In fact, the most brilliant sections of the *Novum Organum*—scarcely hidden under a bushel by Bacon, and well known to subsequent historians, philosophers, and sociologists—refute the Baconian myth by defining and analyzing the mental and social impediments that lie too deeply and ineradicably within us to warrant any ideal of pure objectivism in human psychology or scholarship. Bacon referred to these impediments as "idols," and I would argue that their intrusive inevitability fractures all dichotomous models invoked to separate science from other creative human activities. Bacon should therefore be honored as the primary spokesman for a nondichotomized concept of science as a quintessential human activity, inevitably emerging from the guts of our mental habits and social practices, and inexorably intertwined with foibles of human nature and contingencies of human history—not apart but embedded, yet still operating to advance our general understanding of an external world and therefore to foster our access to "factual truth" under any meaningful definition of such a concept.

The old methods of syllogistic logic, Bacon argues, can only manipulate words and cannot access "things" (that is, objects of the external world) directly:* "Syllogism consists of propositions, propositions of words, and words are the tokens and marks of things." Such indirect access to things might suffice if the mind (and its verbal tools) could express external nature without bias; but we cannot operate with such mechanistic objectivity: "If these same notions of the mind (which are, as it were, the soul of words) . . . be rudely and rashly divorced from things, and roving; not perfectly defined and limited, and also many other ways vicious; all falls to ruin." Thus, Bacon concludes, "we reject demonstration or syllogism, for that it proceeds confusedly; and lets Nature escape our hands."

Rather, Bacon continues, we must find a path to natural knowledge—as we develop the procedure now known as modern science—by joining observation of externalities with scrutiny of internal biases, both mental and social. For this new form of understanding "is extracted . . . not only out of the secret closets of the mind, but out of the very entrails of Nature." As for the penchants and limitations of mind, two major deficiencies of sensory experience impede our understanding of nature: "the guilt of Senses is of two sorts, either it destitutes us, or else deceives us."

---

*I take all quotes from my copy of Gilbert Wats's translation (1674) of Bacon's *Novum Organum,* originally written in Latin for accessibility to all European intellectuals in the only shared language of Bacon's time. In an irony of growth and recursion, this valuable commonality disappeared with the decline of Latin and the rise of nationalism in the eighteenth century. Only now has the notion of an international scientific language gained ground again, almost to the point of effective establishment. This time, however, the choice is English—good for us, and galling for some others!

The first guilt, "destitution," identifies objective limits upon physical ranges of human perception. Many natural objects cannot be observed "either by reason of the subtlety of the entire body, or the minuteness of the parts thereof, or the distance of place, or the slowness, and likewise swiftness of motion."

But the second guilt, "deception," denotes a more active genre of mental limitation defined by internal biases that we impose upon external nature. "The testimony and information of sense," Bacon states, "is ever from the Analogy of Man, and not from the Analogy of the World; and it is an error of dangerous consequence to assert that sense is the measure of things." Bacon, in a striking metaphor once learned by all English schoolchildren but now largely forgotten, called these active biases "idols"—or "the Idolae, wherewith the mind is preoccupate."

Bacon identified four idols and divided them into two major categories, "attracted" and "innate." The attracted idols specify social and ideological biases imposed from without, for they "have slid into men's minds whether by the placits and sects of philosophers, or by depraved laws of demonstrations." Bacon designated these two attracted biases as "idols of the theater" for limitations imposed by old and unfruitful theories that persist as constraining myths ("placits of philosophers"); and, in his most strikingly original conception, "idols of the marketplace," for limitations arising from false modes of reasoning ("depraved laws or demonstrations"), and especially from failures of language to provide words for important ideas and phenomena, for we cannot properly conceptualize what we cannot express. (In a brilliant story titled "Averroes' Search," the celebrated Argentinean writer Jorge Luis Borges, who strongly admired Bacon, described the frustration of this greatest medieval Islamic commentator on Aristotle, as he struggled without success to understand two words central to Aristotle's *Poetics,* but having no conceivable expression in Averroes's own language and culture: comedy and tragedy.)

But if these attracted idols enter our minds from without, the innate idols "inhere in the nature of the intellect." Bacon identified two innate idols at opposite scales of human society: "idols of the cave," representing the peculiarities of each individual's temperament and limitations; and "idols of the tribe," denoting foibles inherent in the very (we would now say "evolved") structure of the human mind. Among these tribal idols of human nature itself, we must prominently include both our legendary difficulty in acknowledging, or even conceiving, the concept of probability and also the motivating theme of this book: our lamentable tendency to taxonomize complex situations as dichotomies of conflicting opposites.

In a key insight, explicitly invoking these idols to dismember the myth of objectivity, Bacon holds that science must inevitably work within our mental foibles and social constraints by marshaling our self-reflective abilities to understand—because we cannot dispel—the idols that always interact with external reality as we try to grasp the nature of things. We might identify, and largely obviate, the theatrical and marketplace idols imposed from without, but we cannot fully dispel the cave and tribal idols emerging from within. The influence of these innate idols can only be reduced by scrutiny and vigilance: "These two first kinds of Idolaes [attracted idols of the theater and market-place] can very hardly; but those latter [innate idols of the cave and tribe], by no means be extirpate [sic]. It remains only that they be disclosed; and that same treacherous faculty of the mind be noted and convinced."

In a striking metaphor, Bacon closes his discussion of idols by describing our scientific quest as an interplay between mental foibles and outside facts, not an objective march to truth—a marriage of our mental propensities with nature's realities, done for purposes of human betterment: "We presume . . . that we have prepared and adorned the bridechamber of the Mind and of the Universe. Now may the vote of the marriage-song be, that from the conjunction, human aids, and a race of inventions may be procreated, as may in some part vanquish and subdue man's miseries and necessities."

I can only express a final hope that the consummation of such a favorable union might not only destroy the barren myth of dichotomy forever, but might also, in the healthy hybridization of mental modes (so long understood and so well practiced in the humanities) with techniques of observation and experiment (so fruitfully exploited by the sciences), yield a bevy of mixed off-spring that would expose the concept of oppositional dichotomy between science and the humanities as a foolish negation of our mental capacities and complexities—a trap no less harmful and restrictive of human potential than our former efforts to keep nonexistent human races both separate and unequal.

6

Reintegration in
Triumphant Maturity

YES, A TIME TO BREAK DOWN AND A TIME TO BUILD UP. CLEARLY WE ARE
now in the Preacher's second stage (Ecclesiastes 3:3), and it strikes me as simply unseemly, not to mention unprofitable, to keep the demolition crew in
high perks and wages when we ought to be hiring architects and masons. A
confrontational attitude toward the contrary claims of Renaissance humanists justifiably characterized the initial rhetoric of modern science in its
seventeenth-century infancy, as this new kid in town struggled to gain some
ground in a grand game of intellectual mumblety-peg (a contest of universal
boyhood, called "land" or "territory" in various boroughs of New York City during my youth, and based on dividing up a specified totality of ground by throwing a pocketknife into the earth and cutting along the line of penetration). But,
for two basic reasons, I see no conceivable justification, other than human narrowness and the weight of "traditional" practice, for continued contention
between science and the humanities: (1) science took charge, triumphantly and
ever so long ago, of the empirically designated and logically allotted share of
land on the big board of our mental lives; and (2) the full board includes

both generous parcels for each of the many mansions of our different pursuits (one for each foxy style), and large tracts of shared space for debates, games, joint presentations, and endless schmoozing on lovely park benches (the careful and ever so fruitful joining advocated at the very end of my preface).

Yet science has shown a regrettable tendency both to claim superiority (or at least privileged status) as a "better" way of knowing in general, and also to engage in forays and poachings into mansions that, by elementary courtesy, require an explicit invitation for entrance as someone else's guest. Scientists have tended to depict their own history as a steady march to truth, mediated by successful application of a universal and unchanging "scientific method" that only requires time to clear away the encumbering myths of a "bad old" past bound by strictures of theology or some other social impediment, and to accumulate the empirical data required to validate nature's true modes of operation.

This privileged view of a "separate" science, chugging along progressively while other institutions tack to the ever-changing winds of social fashion, achieved its "purest" expression—thus winning the deserved enmity of intellectual historians who understood the true complexity and contingency of all human disciplines—in the "positivist" philosophy and historiography of the late-nineteenth-century German physicist Ernst Mach. This basic approach, so arrogant in its claims for special status among institutional histories, and so justly rejected by virtually all modern historians of science (see Thomas Kuhn, *The Structure of Scientific Revolutions,* 1962, and Norwood Russell Hanson, *Patterns of Discovery,* 1958, for the classic statements), still motivates the pervasive and ordinary, if uncritical, beliefs of most working scientists about the history of their disciplines, and still festoons the obligatory introductory page about past worthies in virtually every undergraduate textbook in science.

Serious historians dismiss this cardboard version of history as linearly accumulating progress with an odd term of jargon derived not from science at all, but from a group of scholars who plied their craft by using the past to validate Whig principles of their own political affiliation. The great British historian Herbert Butterfield designated this attitude as Whig or Whiggish history in a famous essay (of novella length) published in 1931, in print ever since, and titled *The Whig Interpretation of History.* Although Butterfield took his name from a particular group of political historians, he recognized that this style of presentation had always been favored by the few professional scientists who took an interest in their subject's past. In his preface, Butterfield defined this self-serving approach as "the tendency in many historians to write on the side of Protestants and Whigs, to praise revolutions provided they have been successful, to emphasize certain principles of progress in the past and to produce a story which is the ratification, if not the glorification of the present."

I criticize this triumphalist conviction of most scientists for two primary reasons: First, because Whiggish assertions alienate colleagues in other fields by claiming a special privilege for science and its history as a pristine and progressive form of human knowledge. And, second, in a curious irony, because the antidote to the cardinal Whiggish assumption that scientists free themselves from surrounding social and psychological norms to follow the straight and narrow path toward truth emanated from one of our own, and at the beginning of our modernity: Francis Bacon's classification and analysis of the "idols" of our social and cognitive prejudices (as just discussed in chapter 5, pages 111–112).

Scientists will not make their appropriate and harmonious peace with colleagues in other disciplines until they recognize their own calling as a quintessentially human enterprise, laden with all the mental idiosyncrasies of the species that must do the work, yet still capable, as its own special feature (for every discipline can claim *some* interesting uniqueness) of reaching a more adequate and deeper understanding of material reality.

But I have no desire to pursue this familiar argument against scientific Whiggery any further in such an abstract and non-operational form—for we scientists tend to be suspicious, and rightly so, of grandiose and general claims without any immediate operational oomph. Rather, the argument becomes much more persuasive if we can show our fellow scientists that abandoning the objectivist mythology of Whig history, and acknowledging (perhaps even embracing) the human foibles and social embeddedness in all scientific activity, will greatly improve the daily practice of our own trade. Thus, and returning to the seventeenth-century works of Grew and Ray to illustrate the inevitable and pervasive human side of putative "objectivity" (see previous discussions of these works in chapter 3), I shall explore the two eminently practical reasons why scientists should devote substantial respect and attention to this humanizing character of all our work: (1) significant gains in our own understanding, or becoming a fox to be a better hedgehog, and (2) important melioration of the fears and misapprehensions of a suspicious public, whose approbation we require in a democratic system with scientific research so dependent upon government funding—or showing the skeptics, in other words, that hedgehogs really can be useful and cooperative, despite those overt prickles.

1. An understanding of the social embeddedness of all aspects of science can forge an essential tie with humanistic studies and greatly aid the technical work of scientists as well.

The most harmful effect of objectivist mythology arises from its insidious role (in the technical rather than moral sense) in shielding scientists from recognizing their own biases. In most fields, scholars understand that no person is an island, and that the bell of universal folly tolls for all of us. But most scientists actually believe their own cant, trusting that the "scientific method" frees them from strictures of unconscious preferences for certain social outcomes, cognitive styles, or psychological stances. Thus the loudest scholarly apostles of obedience to factual reality become, ironically, the most gullible prey to subjective biases, lulled into complacency by a belief that their canonical procedures build shields against such impediments. But just as vigilance becomes the eternal price of liberty in our political slogans, so too must rigorous self-scrutiny represent the cost of fairness and maximal objectivity in scientific research. And we scientists can best appreciate both the general principle itself, and the major snares of specific biases, by reading and respecting our colleagues in the humanities and social sciences, the main disciplinary "homes" for study of this ineluctable human side to all forms and styles of inquiry.

From the Socratic injunction *gnothi seauton* (know thyself), to the admonition "physician heal thyself," to Jesus' suggestion that the nonexistent sinless might cast the first stone, our motto-makers have understood the essential principle and paradox of self-awareness as the most difficult (albeit closest) form of obtainable knowledge. Thus, because we experience such trouble in identifying and expunging our own biases—for we so often misequate them with logically evident or factually proven truth, if we recognize them as arising within ourselves at all—the historical study of distant forebears offers maximal insight into the Baconian idols that stand before nature in our struggles to understand this wondrously complex universe. For when we expose the "obvious" social influences so casually depicted as evident factual reality by people whom we admire as undeniably ahead of us in raw brain power, then we should be ready to admit the truly unavoidable, if painful, inference that we too must be wallowing in unrecognized assumptions that future generations will deem just as risible.

The following comment from Grew's dedication to the Royal Society exposes, without a glimmer of recognition, the oldest and most pervasive prejudice of gender (while also displaying, on an even more general plane, the primary cognitive bias behind all our taxonomic schemes—dichotomy itself, with the usual judgment of more and less worthy imposed over the geometry of simple division). Just as Bacon, following a tradition of centuries, depicted nature as passively female and the virility of developing science as actively male in seeking to know her (and Bacon did not shy from a torrent of metaphors about ravishing and possessing, or "knowing" the formerly virginal Miss Nature in a biblical sense), Grew praises a wealthy patron of the Royal Society for putting his land (part of nature's bounty) to scholarly benefit, and for using some proceeds to publish Grew's catalog of the society's collections:

> I have made this address not only to do you right, but to do right unto virtue itself. And that having proposed your exemplary prudence unto others, they may, from you, learn to use the redundant part of their Estates either to a charitable end, as this City will witness for yourself, or the promotion of masculine studies, as in the present case.

Since I view the language police as usually and basically silly in any case, and since Grew's words certainly fall beyond any imaginable statute of limitations, I attach no depth of meaning to this particular comment, but merely record the apparent automaticity and extent of such valuation by gender. But when we come to a more subtle and far more extensive example in Ray's *Ornithology*, we can readily gauge the deep influence of the idol of dichotomy on actual practice—for Ray here embraces this bias (unconsciously, I must assume, despite the egregious ring of his words to modern ears) as a supposedly objective basis for achieving the primary goal of natural history: the development of a factually accurate taxonomy, or classification of organisms.

Ray, as previously discussed (see pages 43–47), structures his argument as a pointed refutation, emerging from new methodologies promoted by the Scientific Revolution, of the basic procedure followed by Renaissance scholars in presenting the materials of natural history: the composition of compendia, with completeness (of all stated opinions and impressions throughout recorded history) rather than discrimination and factual accuracy as the primary goal, and with emphasis on classical claims and sources as the wellspring and full guardian of all knowledge. Ray, in breaking with the encyclopedic tradition of Gesner and Aldrovandi, pledged to rely upon discrimination and elimination, thus including only the hedgehog's kernel of verifiable accuracy. But what claims, and which organisms, should be presented, and which omitted? In particular, what specimen (or set of specimens) shall represent a species, for the engraver cannot draw every variation among the multifarious representatives of each kind. Ray's list of four criteria could not be more revealing, or more illustrative of evident biases among other potential choices, whereas Ray, by his own lights (I must assume), merely reported a decision that seemed to him so objectively mandated that he needed to supply no defense beyond simple affirmation:*

> It is requisite now that we inform the reader what compendious ways we sought to avoid unnecessary expenses with graving of figures:
>
> 1. Of the same species of bird when more figures than one occurred either in divers authors, or our own papers, or both, we caused only one, which we judged to be the best, to be engraven.
> 2. We have for the most part contented ourselves with the figure of one sex only, and that the male.

---

*Lest we doubt that deeply rooted, or virtually unquestioned, assumptions about the structure of social practices or human relationships can change with alarming speed and unanticipated directionality (in ways that we either welcome or deplore), who in my generation would ever have imagined that smoking would shift from the chief public and "harmless" pleasure of more than half the adult world to the leading social vice of our time. Have you ever noticed how, in movies form the 1930s, at least half the cast puffs away at any given moment. Folks my age will also remember—while our children will scarcely credit the claim—that small packs of cigarettes used to arrive gratis on every airplane meal tray. And one dared not ask the passenger in the adjacent seat to refrain from lighting up, for such a request, unless apologetically backed up with a doctor's note about your peculiar respiratory problems, would have ranked as the height of chutzpah or impolite assault upon inalienable democratic rights. To illustrate our improvement on an even more important social question (leaving ever so much room for further melioration), I well remember the words of my egalitarian father when, in 1950 or so, I saw a young mixed-race couple walking hand in hand down the streets of Manhattan (scarcely a segregationist bailiwick), and I gawked at a sight I had never seen before: "Steve, don't feel guilty that you stared in surprise. Some day, as the world improves, such a mixture will seem no more strange than a romantic pairing of a blonde and a brunette." My dad, something of a Pollyanna, rarely predicted social change with accuracy, but I am glad that he made a correct forecast in this case.

3. We have omitted all such dubious icons as we knew not whether they were of true birds or not, or could not certainly determine of what species they were.

4. Of such as differ only in bigness, or if otherwise in such accidents as cannot be expressed in sculpture, we have given only the figure of the greater [that is, the larger in body size].

I can accept point one as a worthy generality (pending the definition of "best"). I certainly approve point three as Ray's commitment to the observational ideals of the Scientific Revolution over the promiscuity of previous compendiasts. But what can we make of the points two and four beyond the stubborn persistence of boyness and bigness as undefined preferences recording the state of being (or at least the aspirations) of the writers themselves?

Proceeding further into the heart of Ray's general method for classifying organisms, we encounter a full system of social judgments embedded within the basic structure of his taxonomy (but probably conceptualized by Ray, if he considered the issue in any explicit way at all, as logically necessary decisions implied by objective facts of biology). Ray organizes his categories of birds by a conventional device still widely used, particularly in handbooks for practical identification: the dichotomous key. These keys, although generally drawn as moving from left to right rather than from the ground to the skies, follow the geometric order of a branching tree, with the largest inclusive category at the base (the left-hand side of a key, corresponding to the central trunk of a tree), with finer distinctions recorded by successive divisions of larger units into two smaller units at each point of branching.

Interestingly, the basic geometry of branching precludes any judgment about relative status of the two resulting units. When a trunk splits evenly, neither branch can claim a preferred position, for the point of bifurcation acts as a pivot, and the two resulting branches can rotate freely around this pivot into any accessible position. Thus, when we draw one branch on the left and the other on the right (for a tree growing upward), or when we depict one branch on top and the other on the bottom (for a system, like Ray's, that bifurcates from left to right), we only follow an arbitrary convention. Left and right, top and bottom, can always be interchanged without altering the topology of the system at all.

Yet, in a social convention that may represent more than a cultural accident of Western life, and may well record some hard-wiring about hierarchies and dominations, we tend to equate big and up with power and righteousness, and down and small with lesser merit. In this context, Ray's key (figure 19) for land birds advertises the potency of unconscious bias in a supposedly objective mapping of nature's factuality. Note how, at every branching point, Ray places the socially favored of the two categories on top and the less worthy, in subjective human judgment, on the bottom—even though, as explained just above, the relative placement of the two branches does not affect the geometry of the system.

The first division places carnivorous birds like the noble eagle above, and frugivorous relatives like the blathering parrots below. The lower frugivore branch then trifurcates by body size into big, medium, and little, so arrayed from the preferred top to the subservient bottom. Meanwhile, the carnivores above undergo a basic division into worthy denizens of daylight above, and stealthy creatures of the night below. The nocturnal category then splits into haves and have-nots, with horned owls above and non-horned owls below (figure 20). The blessed day-fliers, meanwhile, divide by size into "greater" above and "lesser" below. The disenfranchised smaller birds then split again, but this time by explicit human judgment of worth into "more generous" above ("want to be reclaimed and manned for fowling") and "more cowardly and sluggish, or else indocile" below, and "therefore by our falconers neglected and permitted to live at large." (I can't help remarking that current sensibilities would probably grant higher status to these indocile forms for their wit in avoiding human servitude.) These less honored indocile forms then split again, this time by size, into greater above and lesser below. Finally, and introducing the new twist of favored geography for a final division, the lesser indociles split into preferred denizens of Europe above and less well regarded "exotics" below.

Moving up to the "more generous" category of lesser diurnal forms, a final division, citing the venerable principle of "more is better" in purest form, prefers the "long wing'd" above to the "short wing'd" below. Meanwhile, the truly topmost category of large, diurnal, and rapacious birds undergoes its final split to produce an overall "winner"—as the "more generous" eagles above vanquish the "cowardly and sluggish" vultures below. God bless America, and look out for them buzzards.

Moving from the specific social prejudices in Ray's key (based on size, sex, location, and gaudiness) into deeper biases within any such supposedly objective account, we must also consider the very choice of dichotomy itself as a

Figure 19.

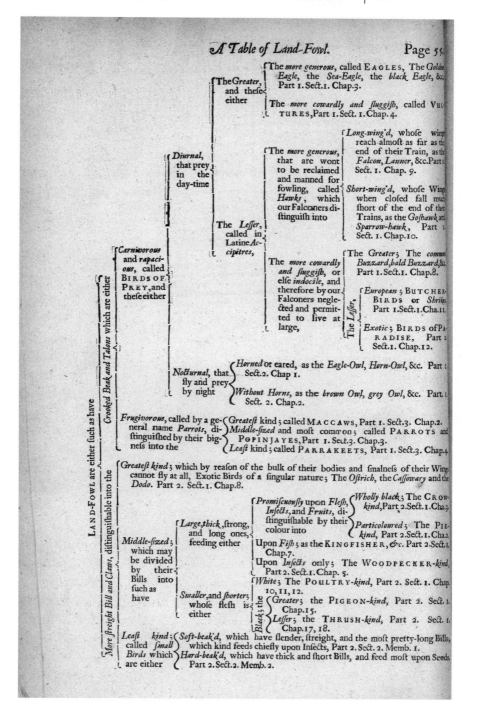

A Table of Land-Fowl. Page 55.

The *more generous*, called EAGLES, The Golden Eagle, the *Sea-Eagle*, the *black Eagle*, &c. Part 1. Sect.1. Chap.3.

The *more cowardly and sluggish*, called VULTURES, Part 1. Sect. 1. Chap. 4.

Long-wing'd, whose wings reach almost as far as the end of their Train, as the Falcon, Lanner, &c. Part 1. Sect. 1. Chap. 9.

Short-wing'd, whose Wings when closed fall much short of the end of their Trains, as the *Goshawk* and Sparrow-hawk, Part 1. Sect. 1. Chap.10.

The *more generous*, that are wont to be reclaimed and manned for fowling, called *Hawks*, which our Falconers distinguish into

The *Greater*; The common Buzzard, bald Buzzard, &c. Part 1. Sect.1. Chap.8.

European; BUTCHER-BIRDS or *Shrikes*, Part 1. Sect.1. Chap.11.

Exotic; BIRDS of PARADISE, Part 1 Sect.1. Chap.12.

The *more cowardly and sluggish*, or else *indocile*, and therefore by our Falconers neglected and permitted to live at large,

The *Lesser*

The *Lesser*, called in Latine *Accipitres*,

Diurnal, that prey in the day-time

Horned or eared, as the *Eagle-Owl*, *Horn-Owl*, &c. Part 1. Sect.2. Chap 1.

Without Horns, as the *brown Owl*, *grey Owl*, &c. Part 1. Sect. 2. Chap.2.

Nocturnal, that fly and prey by night

Carnivorous and rapacious, called BIRDS OF PREY, and these either

Greatest kind; called MACCAWS, Part 1. Sect.3. Chap.2.

Middle-sized and most common; called PARROTS and POPINJAYES, Part 1. Sect.3. Chap.3.

Least kind; called PARRAKEETS, Part 1. Sect.3. Chap.4.

Frugivorous, called by a general name *Parrots*, distinguished by their bigness into the

Greatest kind; which by reason of the bulk of their bodies and smalness of their Wings cannot fly at all, Exotic Birds of a singular nature; The *Ostrich*, the *Cassowary* and the *Dodo*. Part 2. Sect. 1. Chap.8.

Wholly black; The CROW-kind, Part 2. Sect. 1. Cha.3.

Particoloured; The PIE-kind, Part 2. Sect.1. Cha.2.

Promiscuously upon *Flesh*, *Insects*, and *Fruits*, distinguishable by their colour into

Upon *Fish*; as the KINGFISHER, &c. Part 2. Sect.1. Chap.7.

Upon *Insects* only; The WOODPECKER-kind, Part 2. Sect.1. Chap. 5.

Large, thick, strong, and long ones, feeding either

White; The POULTRY-kind, Part 2. Sect. 1. Chap. 10, 11, 12.

Greater; the PIGEON-kind, Part 2. Sect. 1. Chap.15.

Lesser; the THRUSH-kind, Part 2. Sect. 1. Chap.17, 18.

Smaller, and shorter; whose flesh is either

Black; the

Middle-sized; which may be divided by their Bills into such as have

Soft-beak'd, which have slender, streight, and the most pretty-long Bills; which kind feeds chiefly upon Insects, Part 2. Sect. 2. Memb. 1.

Hard-beak'd, which have thick and short Bills, and feed most upon Seeds. Part 2. Sect.2. Memb. 2.

Least kind; called *small Birds* which are either

LAND-FOWL are either such as have

Crooked Beak and Talons which are either

More streight Bill and Claws, distinguishable into the

basis of division—one of the primary "tribal idols" in Bacon's analysis of our cognitive preferences. Much of the world comes to us as continua, or as other complex, and far more than two-valued, series of reasonably discrete states. We do construct useful simplifications when we force this complexity into a simple system of successive dichotomous branchings—for such sequential ordering does resonate with our mind's capacity to grasp a structure within multifarious and hierarchical systems. But what truer or more insightful ways of classification do we miss when we invoke this almost automatic mental scheme without pressing ourselves to consider less congenial, but perhaps more rewarding, alternatives?

Figure 20.

Oddly enough, and to expose my own parochialism, I used to think that dichotomous keying had been invented by biologists for displaying Linnaean systems in the clearest possible way. After all, I learned the rules and techniques for constructing dichotomous keys in my basic undergraduate biology course, and, as mentioned above, botanists and zoologists have been using such keys for centuries. But, in fact, dichotomous keying represents one of our oldest and most general cognitive inventions, used for ages, and across all disciplines, for organizing complex systems of information. In fact, centuries before anyone ever thought about an empirically based classification of organisms, medieval schoolmen, following Saint Thomas and Aristotelian logic, used dichotomous keys as their primary device for displaying the conceptual structure of any classification. (And since classification is the primary Aristotelian technique for understanding causes, this scheme of analysis attains a maximal generality of potential use.)

For example, in 1586, a century before Ray constructed his dichotomous key for the classification of birds, the French jurist Nicholas Abraham published a schoolboy's textbook on logic and ethics. He presented the guts of his classification as a strictly dichotomous key, exactly in the same form that Ray would use one hundred years later.* He also follows Ray's (presumably unconscious) practice of placing the favored category above the less distinguished choice at each division (figure 21)—even though the basic geometry of such branches cannot specify any "above" or "below." For his primary separation, Abraham divides the basis of ethical judgments into *Mentis* (of the mind) above and *Moris* (by custom) below, for rational decisions trump social conventions. Figure 21 then shows subsequent divisions of the preferred mental category. Reasons of the mind divide into *Sapientia* (by wisdom) above and *Prudentia* (by discretion) below, as reason beats out emotion or convenience. *Sapientia* makes a final division into *Intelligentia* (by understanding) above and *Scientia* (by knowledge) below, as a well-wrought abstract argument triumphs over a factually forced decision. *Prudentia* then undergoes its final division into *Bono consulatio* (by good deliberation) above and *Sagacitas* (by the acuteness of personal decision) below, as the agreement of a collectivity trumps the uncertainty of a personal judgment, however wise the individual.

---

*Biologists, in their narrowness, often think that their inventions, so cleverly contrived, have then spread to other disciplines—whereas, in reality, we have usurped someone else's innovation or terminology. In my favorite example, Linnaeus awarded the name Primates—meaning "first" in Latin—to the mammalian order of monkeys, apes, and humans, in obvious reference to their superior mentality. Since all biologists regard this term as our property, we become amused when we encounter the ecclesiastical usage of "primate" as the chief bishop among all others in a nation or larger region—for we can only conjure up an image of a holy man with a miter, scratching himself on all fours in a cage at the local zoo. But the church rightly owns the term by priority of several centuries over Linnaeus's borrowing. At least some ecclesiastics keep a good sense of humor about our usurpation. A wonderful letter from the spokesperson of the Primate (chief Anglican bishop) of Canada to the hapless compilers of a questionnaire about monkeys and apes at local zoos has been copied and widely circulated among biologists throughout the world. The spokesperson reports that his reverend boss is not in captivity, does not care for bananas, and lives in a house with his wife and kids.

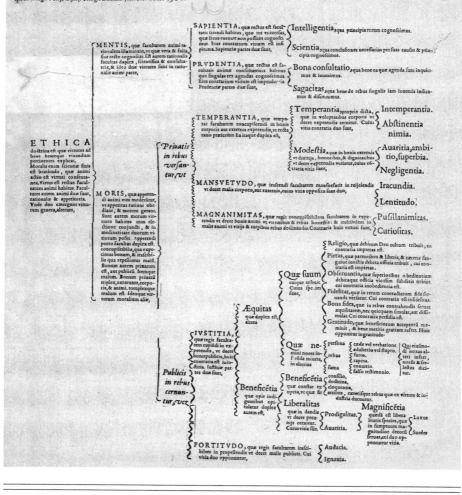

Figure 21.

But to show the depth and venerability of this primal idol of our tribe, consider the following example from the very heart of medieval scholastic traditions, as revealed in a marginal annotation in the oldest book of my personal collection. I own a lovely copy of Saint Thomas Aquinas's commentaries on Aristotle—about as canonical as you can get for the genre, with the greatest of all medieval scholars explicating the greatest of all classical gurus—published in Cologne in 1487 or 1488, just a generation after Gutenberg's invention of the printing press.

If I may venture an almost embarrassingly gushy personal comment, I can hardly begin to explain my pleasure in studying this text from the earliest days of the greatest invention in the history of intellectual life. (Book collectors refer to all volumes published before 1500 as "incunabulae," literally, from the cradle.) For my copy has been so extensively annotated, in an ancient style of handwriting that cannot differ much from the date of the publication itself, that the inked additions often double the total verbiage. I do not have a particularly active imagination, and therefore fail to enjoy the pleasures felt by some people who can observe the ruin of a single column from a classical Greek temple, and then conjure up, in their mind's eye, not only the entire edifice, but also the practices and feelings of the original utilizers and inhabitants. But, somehow, I can so proceed with the annotations in this book because they are so extensive, written *in schvartz* (or sometimes in the red ink of rubrication), and therefore preserved in their totality. I can see a late-fifteenth-century consumer, a university student or an aspiring cleric perhaps, sitting by his nighttime candle (for this book preserves wax drippings on several pages), trying to puzzle out the logic of the great masters, and quickly writing down his epitomes, lest he forget.

The diligent annotator of Aristotle's *De Anima* included several branching keys among his marginalia. They do not always follow a purely dichotomous pattern, as some divide their principal subjects into three or five subcategories. But strictly dichotomous keys, moving from left to right just as Abraham's for ethics in 1586 and Ray's for birds in 1678, abound. One example, written into book two of Saint Thomas's commentary on Aristotle's *De Anima,* intrigued me to the point of virtually proving a case for strong human inclinations toward sequential dichotomy, or successive divisions into pairs, as a preferred mental device for classifying complex systems. For, in this case, Aristotle's text suggests either a single continuum with three ordered categories, or a dichotomous tree with a primary division of two, and a second division of just one of these subcategories into two further groups, again for a total of three, as Aristotle's formulation clearly requires. The annotator of my copy opted for the dichotomous tree with two divisions as his device for generating the three categories.

Aristotle here discusses the various categories of our *intellectus,* or understanding. Saint Thomas points out that our understanding can manifest itself either *ad actu* (by action or impulse) or *potentia* (by potential).* He then focuses his discussion on the modes by which mere potential can lead to action *(ducit de pona ad actu).* Saint Thomas first presents the single continuum in three stages: *intell[e]c[t]us e[st] in triplici dispo[sition]e* (the intellect is arrayed into three categories). (I have resupplied the missing letters in brackets, as the printer relies heavily on abbreviations as explained in the foregoing footnote.) But he then, immediately thereafter, suggests the alternate classification by dichotomy, with two splits. After the primary division into action and potential, the category of action remains discrete and divides no further. But the second category of *pona (potentia,* or potential) must then undergo a second dichotomous split: *ille mod[us] s[u]bdivit i[n] duos* (this mode divides into two), with a more easily activated category called *propinqua,* or nearby; and a less mobilizable division called *remota,* or remote.

Thus we may conceptualize the entire system either as a continuum of three states from most immediate to most distant (direct action, potential for easily inspired action, and potential more difficult to mobilize); or as a double dichotomous division, first into action and potential, with action then undivided, and potential further split into greater and lesser propensity for recruitment. Faced with these clear choices, probably the two most fundamental alternatives within our cognitive capacities (a single smooth continuum versus a set of successive dichotomous divisions), note how our diligent

---

*At this early stage in the history of printing, publishers had not yet fully recognized the advantages and transforming power of movable type. This printed book, from 1487, still uses cryptic and extensive abbreviations for many words, converting the entire text into a form of shorthand. These conventional abbreviations had greatly boosted the speed of production for texts, when each copy had to be written out by hand, but saved little time, and perhaps a little more space—but only at the cost of great ambiguity and difficulty in reading—when the type for each word only needed to be set once. Thus these abbreviations slowly faded from use, leading to our modern conventions of writing texts in full. But the old abbreviations still prevail, though set in type, in this book from 1487. Thus, for example, the word *potentia* becomes *pona* both in the printed type and in my reader's marginal annotations.

student and writer of marginalia opts for dichotomy. In a beautiful hand, our annotator has made his choice and drawn a dichotomous key, moving from left to right through two divisions, first into *actu* above (by action) and *pona* below (*potentia*, or by potential), with *pona* subdivided into *ppinq* (*propinqua*, or nearby, in the sense of easily activated) and *remota* (or distant, in the sense of disinclined).

As a little footnote too sweet to omit (although irrelevant to the subject of classification by dichotomy), our diligent student shows his humanity in a conventional manner across the centuries, from his Latin to our barhopping. As a final comment to his discussion, Saint Thomas notes that easily inspired actions in the category of *propinqua* can be suppressed by circumstances external to the character of the intellect itself. He specifically mentions two: *dolor vel ebrietas* (sadness or drunkenness). Our commentator dutifully records one of these impediments in his elegant hand on the very next page: *Ebrietas impediat scientia[m]* (drunkenness impedes knowledge). I only mention this note because, at the very end of the book, after the printed AMEN and several hundred pages of copious annotations, our student finally grants himself a break, medieval style, by writing: *Claudite jam rimos pueri sat prata biberunt,* or (roughly): "Enough, boys, now let us conclude these thorough investigations and go to meadows to drink." So much for intellectual potential, at least for a celebratory conclusion!

To cite a closing example, from the celebrated French physician and surgeon Ambroise Paré (1510–1590), and illustrating a case where the tribal idol of dichotomy imposed a false solution that impeded the development of medicine for centuries, the tentacles of this cognitive bias extended far beyond simple branchings in sequential keys. More-complex classifications could also be constructed by allowing several dichotomous alternatives to interpenetrate, and by then listing all possible permutations to enumerate the potential categories. The old medical theory of four bodily humors may not, at first, reveal any dichotomous basis. But, in fact, this fallacious system roots its taxonomy in the intersections of two dichotomous divisions between hot versus cold, and wet versus dry.

Under this theory, bodily health requires a balance of four distinct principles or humors (literally liquids)—blood, phlegm, choler, and melancholy. An emphasis upon one of the four leads, successively, to distinctive temperaments or styles of personality that continue, at least descriptively, to designate certain human propensities: sanguine, phlegmatic, choleric, and melancholic. If the humors get more seriously out of whack, then bodily malfunction will ensue—not from the invasion of any foreign agent (as in the later germ theory of disease) or from failure to take in essential nutrients (at least not immediately, but only through their influence on the production of humors), but directly from the internal imbalance itself. The remedy for disease, with illness thus construed as an imbalance among humors, must focus upon techniques for reducing overactive humors and restoring the weakened components. The theory of humors therefore inspired centuries of belief in a large array of procedures now regarded as entirely ineffective, if not barbaric—including bloodletting (to reduce the sanguine humor), sweating, purging, vomiting, et cetera.

But why, in the absence of any direct evidence for the existence of such liquids, did classical medicine insist so strongly upon four, and only four, humors? The standard solution, so congenial to two modes of thought that flourished before the Scientific Revolution and then died with its success, invoked, first, a correspondence between the microcosm of the human body and the macrocosm of the universe; and, second, the four permitted categories of a double dichotomy to define the corresponding divisions of both the microcosm and macrocosm. Just as four humors balanced the microcosm (blood, phlegm, choler, and melancholy), so too did four elements (air, water, fire, and earth) build the macrocosm. In each case, these four represent all possible combinations of the two primal dichotomies for material things: hot versus cold, and wet versus dry.

Paré's chart (from my 1614 edition of his collected works, figures 22 and 23) lays out all aspects of this system explicitly: blood corresponds with air and represents the hot and wet substance; phlegm represents water, the cold and wet element; choler corresponds with fire, hot and dry; whereas melancholy represents earth, cold and dry. (Note how the temperaments arise from this conception, with the hot and wet person as sanguine, or optimistic and level-headed; the cold and wet person as phlegmatic, or slow to act; the hot and dry person as choleric, or quick to anger; and the cold and dry person as melancholy, or just plain sad.)

I have, in this prolonged discussion, only addressed one helpful insight (albeit the primary benefit, in my judgment) that scientists could gain from their colleagues in the humanities: exposing the myth of objectivity by a positive acknowledgment (not a cynical and despairing shrug for inevitable loss) of the mental quirks and social influences upon all factual study of the natural world—for honest recognition can only breed self-awareness and greater

LES
OEVVRES
D'AMBROISE PARÉ
CONSEILER, ET PREMIER
CHIRVRGIEN DV ROY.

CORRIGEES ET AVG-
mentees par luy-mesme, peu
au parauant son decés.

Diuisees en vingt-neuf liures.

Auec les figures & portraicts, tant de l'Ana-
tomie que des instruments de Chirur-
gie, & de plusieurs Monstres.

SEPTIESME EDITION.

Reueüe & augmentee en
diuers endroicts.

A PARIS,
Chez BARTHELEMY MACÉ,
au mont S. Hilaire, à l'Escu
de Bretaigne.
1614.
Auec Priuilege du Roy.

Figure 22.

A

	Nature.	Côsistéce.	Couleur.	Saueur.	Viage.
Le sang.	De la nature de l'air chaud & humide ou pluſtoſt temperé.	Mediocre, ny trop eſpais ny trop clair.	Rouge & vermeil.	Doux.	Il nourrit principalement les parties muſculeuſes : eſt diſtribué par les veines & arteres, donne chaleur à tout le corps.
Le phlegme ou pituite.	De la nature de l'eau, froide & humide.	Fluxile.	Blanche.	Douce ou pluſtoſt fade : car ainſi eſtimons-nous ceſte eau bonne qui n'a aucun gouſt.	Elle nourrit le cerueau, comme auſſi toutes autres parties froides & humides : modere le ſang, & aide le mouuement des articles.
La cholere.	De la nature du feu, chaude & ſeiche.	Ténuë & ſubtile.	Iaune ou paſle.	Amere.	Elle excite la vertu expultrice des inteſtins, atténuë le phlegme qui eſt en iceux: ce que i'entends de l'excremĕtitielle: cõme auſſi l'alimentaire nourrit les parties qui approchĕt plus pres de ſon naturel.
L'humeur melancholic.	De la nature de la terre, froid & ſec.	Gras, eſpais, & limoneux.	Noir.	Acide & poignant.	Il excite l'appetit, il nourrit la ratte, & toute autre partie, qui luy eſt ſemblable en temperature, comme les os.

C

Le ſang eſt fait de la partie la plus benigne de tout le chylus, contenu és veines, & *Dequoy &* principalement eſt formé au foye, ainſi qu'auons dit: il eſt procreé des alimens de *en quel* bon ſuc, prins apres exercices moderez : & plus en vn aage qu'en vn autre : & en vne *temps ſe* partie de l'année conuenable plus qu'en l'autre, qui eſt le Printemps, lequel du tout *fait le bon* approche à la nature du ſang : (dont f'enſuit que le ſang ſoit temperé en ſes qualitez, *ſang.* non chaud & humide, cõme ainſi ſoit que ſelon l'opinion de Galien au premier des *Confirma-* Temperamens, le Printemps eſt auſſi temperé, comme a eſté touché par cy-deuant.) *tion de la* Parquoy en ce temps ſont faites couſtumierement les bonnes ſaignées. L'aage fort *temperatu-* propre à engendrer tel humeur eſt l'adoleſcence, ou comme dit Galien, depuis *re du ſang.* vingt-cinq iuſques à trente-cinq : ceux, auſquels tel humeur abonde, ſont moderez, rouges, coulourez, amiables & vermeils, ioyeux & plaiſans.

Le phlegme eſt fait des alimens froids & cruds, mais principalement en hyuer & en vieilleſſe, à raiſon de la cõſtitution froide & humide, tãt de l'aage que de telle partie
D de l'an. Il rend l'hõme endormy, pareſſeux & gras, ayant trop toſt les cheueux blãcs.
La cholere eſt comme la fureur des humeurs, laquelle eſt engendrée auec le ſang au foye, & portée és veines & arteres: & celle qui excede, eſt enuoyée en partie au folicule du fiel, en partie f'exhale par inſenſible trãſpiration & ſueurs: car le ſang des arteres eſt plus ſubtil, & plus iaune que celuy des veines, ainſi que dit Galiẽ. En ieuneſſe & en *Au liure* Eſté eſt fait tel humeur, tãt des viãdes acres, ameres ou ſalées, que du trauail d'eſprit *6. de locis* & du corps: auſſi tel humeur eſt principalement purgé en tel temps. Il rend l'homme *affeætis.* leger, ſubit, facile à ſe cholerer, & prompt à toutes choſes, maigre, agile, qui a toſt fait digeſtion des viandes qu'il a priſes. L'humeur melancholique eſt la partie la plus groſſe du ſang, lequel en partie eſt reietté du foye, & attiré par la ratte pour la nutrition d'icelle & expurgation du ſang en partie porté auec le ſang, pour nourrir les parties de noſtre corps les plus terreſtres. Il eſt fait des alimens de gros ſuc & difficiles à cuire, & auſſi des ennuis & faſcheries de l'eſprit : il redonde principalement en Automne, ou en l'aage declinant & premiere vieilleſſe : & rend tel humeur les hommes

b iiij

Figure 23.

practical sophistication about the mental processes that scientists must use to reach their accurate conclusions. But I also wish to mention, more briefly, two additional factors, also better known, appreciated, and more widely studied within the humanities—valuable stratagems of the fox for adding some excellent and fully honorable nuances of real effectiveness to the overly restricted or inadequately examined practices of science.

Humanists, for my second point, rightly stress the virtues and felicities of stylistic writing, not as a mere frill or foppish attribute, but as a primary aid to attention and understanding. Scientists, on the other hand, and as a virtual badge of membership for admission to our professional club, tend to assert that although brevity and clarity should certainly be fostered, the nurturing of verbal style, as an issue of form rather than substance, plays no role in the study of material reality.

In fact, this explicit denial of importance to modes of communication has, unfortunately, engendered a more than merely mild form of philistinism among many scientists who not only view verbal skills as unimportant, but actually discount any fortuitous stylistic acumen among their colleagues as an irrelevant snare, casting suspicion upon the writer's capacity for objectivity in presenting the data of nature. In an almost perverse manner, inarticulateness almost becomes a virtue as a collateral sign of proper attention to nature's raw empirics versus distilled human presentation thereof. (And yet, to cite a pair of ironies, proving that the best scientists have always understood the value of both assiduous data gathering and elegant communication, John Ray composed even his denial of the importance of good writing—see the quotation on page 47—in his characteristically excellent prose. And the famous motto *"le style c'est l'homme même"* (style makes the man) did not emanate from a leader among the literati, but from the finest naturalist of eighteenth-century France, and a great writer as well—Georges Leclerc Buffon, whose forty-four-volume *Histoire naturelle,* equally admired for both style and content, became the first great encyclopedia of modern approaches to the study of nature.

Because we have cut ourselves off from scholars in the humanities who pay closer attention to modes of communication, we have spun our own self-referential wheels and developed artificial standards and rules of writing that virtually guarantee the unreadability of scientific articles outside the clubhouse. Some of our conventions might also be called ludicrous in their utter failure to achieve a stated end, and in the guaranteed clunkiness of style thus engendered by rules that any good writer would immediately recognize as crippling. In my favorite example, scientists have trained themselves to write in the most unfelicitous of all English modes: the unrelenting passive voice. If you ask scientists for a rationale, they will reply with the two standard defenses: economy of presentation and objectivity of statement. Neither, in fact, makes any sense. Sentences in the passive voice tend to be longer than the corresponding active statement, while immodesty and personal glorification can proceed just as readily without the dreaded "I." Which of the following do you prefer for brevity, modesty, and just plain felicity: "The discovery that was made was no doubt the most significant advance of our times"; or "I have discovered a procedure to solve the persistent problem . . ."?

Sometimes, at least, our unquestioned dedication to such literary barbarisms can yield some humor to lighten a tough day at the office. I once, for example, blue-penciled the following comment from the dissertation of an earnest graduate student, committed to following the rules and joining the club, but untutored in the fundamentals of English prose. He wished to make the important point that his elaborate measurements of human skulls required more than a morning's work, and that he had to insert a break between the two halves of his protocol. He wrote, "The room was then left for lunch"— and I could only imagine the office furniture grabbing a quick roast beef sandwich when the human occupants took a break at midday.

This lack of attention to style, combined with an active belief that quality of prose cannot impact the power of an argument, at least confers an admittedly undeserved blessing upon those few scientists who, by rare training or good fortune, happen to write unusually well and persuasively. In the humanities, such verbal power would be recognized and properly discounted in judging the logical acumen of any argument. But scientists believe that only the quality of data and the logic of presentation pack persuasive punch, and simply do not recognize—and can therefore be invisibly influenced by—the sheer power of prose, even in support of a dubious case. To cite my two favorite examples of great writing that carried questionable arguments, Charles Lyell became the father of geology, and the apostle of gradualistic change, more by the extraordinary quality of his elegant and lucid prose (in the three volumes of his *Principles of Geology*, published between 1830 and 1833) than for the evident veracity of his theories of change or the quality of his field work. (Lyell's poor eyesight guaranteed that personal observation of strata and landforms would play little role in developing or supporting his views.) Rather, as a trained barrister with a wonderful flair for polemical writing (the primary desideratum of his original day job, after all), Lyell won his case for uniformitarianism more by composing a brilliant brief than by empirical documentation. Then, in our last century, Sigmund Freud rose to preeminence as a paramount social force through his unparalleled literary gifts, and surely not for his cockamamie and unsupported theory of the human psyche. If *The Interpretation of Dreams* had been written in the unrelenting passive voice of more-scientific prose, I doubt that Mr. Freud's theory would have attained the status embodied in the literal meaning of his name—joy.

Given my commitment to reciprocal enlightenment between science and humanities, and not wanting so thoroughly to castigate my own colleagues or my revered profession, let me close this diatribe by pointing out, lest my humanistic readers become smug, that we scientists have also figured out a foxlike thing or two about communication, and that you would do well to heed the rustics and naifs operating by the seat of their pants within the world of science. We may generally write poorly, and by rules of our own construction that make no sense under any ideal of stylistic felicity. But we generally talk ever so much better than you do—and for a pair of reasons related in reverse to our failures in writing: because, in this sphere, unlike our writing, we have not set poor rules for false purposes, whereas you have done so, and have therefore failed by disregarding a natural inclination toward proper communication.

As a trade secret of the academic arts and humanities, scholars in these disciplines almost always read their papers from previously prepared texts. I find this odd procedure counterproductive (a diplomatic and euphemistic term masking the strong labels I might otherwise be tempted to apply) for a host of reasons. Above all—and folks in the humanities who count words as their stock-in-trade should know this principle better than anyone else—written and spoken English are quite separate languages, and should never be thus confused. Written texts are spare, formal, and nonrecursive at best (for a reader can always return to something missed the first time around). Spoken English, by contrast, must employ repetition to reinforce points that have faded into a nonrecoverable temporal void, and must proceed with greater informality, lest a barrier rise to human contact with a face and body directly before one's eyes.

I challenge anyone who denies the difference to read Martin Luther

King's "I Have a Dream" speech, beyond dispute one of the greatest orations of the twentieth or any other century. But the text fails as written English because its poetic repetitions, based on "let freedom ring" and "I have a dream" do not work as silently read prose. For a less potent example, I could never figure out why such doggerel as "Casey at the Bat" became the most famous of American baseball poems—until I heard someone read it out loud, and realized that the piece had been composed for declamation, a common party activity in nineteenth-century drawing rooms, and not for silent reading. The klutzy but perfectly rhyming meter and lines make perfect sense (and drama) in verbal presentation, but not before one's silent eyes *in schvartz.*

For a second reason, most people read aloud very badly, without inflection or emotion, and with eyes downcast on the text. So even if a written text reads well, few people will execute the task adequately. Finally, we must face an almost ethical point for harried academics. Why should I come all the way to attend a talk, just to hear someone read a text poorly in real time that (since the speech already exists in a printed version) I could read for myself, and probably with greater profit, at one-tenth the temporal expense?

While I am pursuing this rant, let me mention my other pet peeve about papers presented at meetings by scholars in the humanities. With the exception of art historians, who, by good custom, always use two slide projectors simultaneously, humanistic scholars almost never show any pictures at all—even for subjects with the most clearly intrinsic visual content. In fact, slide projectors are rarely available at humanistic meetings, even if an alien speaker should come prepared to show some pictures. I have facetiously remarked that if I ever had my name attached to any natural principle, I would specify the following rule as "Gould's Law": If you, as a scientist, are ever asked to give a talk to a humanistic audience, please remember to ask in advance for a slide projector. (Scientists invariably use visuals and know that projectors will always be available. In fact, the opposite sin of innumerable scientific talks lies in our tendency to darken the room immediately when we step up to the podium, thereby sending a high percentage of the audience to dreamland, and then to structure the talk around a continuous series of slides. An old joke among scientists asks: What would Galileo's opening line have been if he had initially presented *Sidereus nuncius*—his revolutionary "pamphlet" reporting his first telescopic observations of the heavens—as a talk at a modern professional meeting? The answer, of course: "First slide, please.")

I will merely describe, as evidence, my oddest academic experience—in

Paris several years ago at a major international conference to celebrate the two hundredth anniversary of the great Natural History Museum. One could scarcely imagine a more visual subject (and primates are visual animals, after all), as speaker after speaker told their tales of specimens in the museums, animals in the adjacent zoo, and curators who once led the world in scholarship. But not one speaker from a department of the humanities showed slides. In fact, only three speakers presented any visual material at all: Martin Rudwick and I, both trained as professional scientists, but now also engaged in academic work in the history of science (Martin as a true professional, I as an informed amateur), and the curator of the collection of wax models at the museum, who could scarcely not show images of the magnificent objects under his care.

I don't know why humanists, the supposed experts and guardians of good language, fail to grasp this elementary point about the difference between written and spoken English. I can only surmise that they so fear the possibility of any slip if they speak spontaneously—a misplaced preposition, God forbid—that they discount a proper intuition about effective presentation and opt for the safety of something fully prepared in advance. (As the Holiday Inn used to advertise: "no surprises.") But we should all heed the wise advice of a true master in all modes of communication, Thomas Henry Huxley, who stated that a speaker could present a paper in three ways, but should nearly always select the third mode as best: (1) impromptu, or in modern parlance, "winging it" without much thought or preparation, which one should never do, if only for the disrespect thus shown to audiences; (2) by reading a text, which one should also generally avoid for all the reasons expressed above; and (3) extemporaneously, or well prepared and thought out beforehand, but then spoken directly in the unwritten language of oral presentation, which, Huxley advises, a good speaker should almost always do. Virtually every scientist speaks extemporaneously all the time. I do not think that we so proceed because we have explicitly developed any proper theory about the differences between spoken and written English, but largely because we value informality (but not carelessness), and would never spend hours writing for ten minutes of reading. I expect that scholars in the humanities fail to heed Huxley's principle because they fear that a truly extemporaneous speech will be mistaken for an impromptu presentation, then rightly castigated. We all need to learn the crucial difference.

My third and final point matches or exceeds the others in importance, but has been discussed before and need only be summarized here: Science may have a unitary goal—to document the factual character of the material world and to explain why nature operates as it does, and not in some other conceivable way; that is, and roughly, to ascertain fact and explicate theory. But nature works in many ways its wonders to perform, and conventional procedures in science do not always resolve these modes in an optimal or most insightful manner—not because "science" itself cannot, in principle, generate the appropriate range in ways of knowing the empirical world, but rather because the contingent history and conventional sociology of science has favored some modes and largely ignored others. In particular, and in a legacy dating back to the Scientific Revolution itself, the practice of Western science has strongly favored quantitative and experimental techniques so brilliantly suited to the resolution of relatively simple systems, causally set by a few determining variables subject to experimental manipulation, and operating under invariant laws of nature that impart no history to a subject's phenomenology, but always operate in predictable ways, under definable circumstances.

Yet a large range of factual subjects, evidently part of science and duly explainable (in principle) by empirical methods operating under natural laws, treats different kinds of inordinately complex and historically contingent systems—the history of continents and landforms, or the pattern of life's phylogeny, for example—as not deducible, or predictable at all, from natural laws tested and applied in laboratory experiments, but crucially dependent upon the unique character of antecedent historical states in a narrative sequence fully subject to explanation after the fact, but unpredictable beforehand. Narrative explanations of this kind could have been developed within the sciences, but were underplayed or ignored in these realms because the particular history of disciplinary specialization in Western universities allocated this way of knowing primarily to historians in departments of the humanities. Our intellectual taxonomy need not have developed in this manner, but it did— and the socially defined institution of "science" therefore failed to nurture or often to understand at all, or, in worst cases, even to reject explicitly as outside its bailiwick and therefore unworthy in principle, several important modes of explanation that regulate many aspects of the empirical world (and therefore become part of science by broad definition and legitimate range of options by the fox's good and flexible approach to workable strategy).

As a scientist who does much of his study in this historical domain— trying to know the reasons for particular incidents and patterns in the unique history of life, as well as pursuing the more conventional scientific goal of try-ing to explicate the timeless generalities of evolutionary theory—I have found the standard techniques of my discipline quite inadequate, and often even misleading, in my quest to understand the nature of causality in contingent historical sequences that can occur but once in all their detailed glory. I have therefore actively sought insights from theorists in the study of human history. In particular, I never grasped the crucial—and eminently knowable—role of contingency in the history of life until I learned why the South had lost the Civil War, not as a predictable and inevitable consequence of superior Northern forces and firepower, but as a contingent result of many particular events, each of which could have unfolded in an opposite direction, but did not for resolvable reasons effectively unrelated to general laws of nature (or even to Voltaire's quip that God always favors the bigger battalions), but cru-cially dependent upon quirks of individual human decision.

In summary, the three themes discussed here should establish a strong case for the practical value—not to mention the abstract beneficence in an ecu-menical and irenic world—offered by humanistic studies as three foxlike strategies of great potential benefit to the *operational* world of science. A more perfect union of our falsely sundered disciplines would offer powerful bene-fits, in terms of insights and methods of study, for furthering the ordinary work of empirical science. In particular, I have here extolled the superior understanding of the humanities in three areas: (1) acknowledging and ana-lyzing the social influences and cognitive biases within and behind all creative work, including empirical studies; (2) emphasizing the importance of stylistic and rhetorical concerns in the presentation and acceptance of any good argu-ment; and (3) developing certain modes of knowing that science needs but, for contingent reasons of its own history, never emphasized or even down-graded, but that flourished instead within the humanities. In short, humanis-tic study can teach scientists to recognize embeddedness, value style, and access additional modes of explanation. Science, in return, offers just as much to the humanities—so reintegration, after so many centuries of mutual suspi-cion and denigration, should rank high on everyone's list of priorities.

2. *A sympathetic application and understanding of "user friendly" themes in humanistic study will aid the approbation and acceptance of science by a suspi-cious general public. The breaking down of artificial barriers between the sciences and humanities will help even more.*

As a flip side to aid offered by humanistic study in extending and refining our own explorations of the natural world, the same themes can help us to bear (and lighten or even discard) our other major cross in this modern world of suspicion and division: the widespread perception of science as an alien and incomprehensible force in contemporary society; and, even more perniciously, the widespread impression that the practice of science somehow confutes the ethical norms of human decency, or even threatens human continuity by its intrinsic procedures and dangerous knowledge.

The insights of the humanities offer a direct exit from the first dilemma of perception as alien. The fascination of science has always won the affections of a substantial percentage among our population. One need only conjure up the familiar image of a kid in his basement with a chemistry set or a microscope. But this picture also includes the seeds of suspicion and limitation—for the kid is a boy, and also a lonely nerd, preferring his own solitude to a game of stickball (soccer, these days, I suppose) with his classroom or neighborhood buddies. We scientists, in fact, have failed notoriously in our responsibility to foster and maintain the interest and approbation of the general public. We have constructed an arcane jargon that makes us look like a hedgehog's impenetrable ball of sharp prickles, thus driving interested, but untutored, people away. And we have fostered the impression of science as a closed priesthood, penetrable only by rigorous study in certain fields—advanced mathematics, in particular—that do not match everyone's abilities or sensibilities, and that scare many otherwise fascinated folks permanently away.

Creative, frontline work in several sciences does require this kind of mathematical training and experimental skill—and not everyone can muster the requisite ability, generate the necessary energy, or win the appropriate access. But just as few of us could ever, no matter how we might practice, learn to play the violin with sufficient skill to win membership in a world-class orchestra. A crucial paradox therefore arises: Why do we regard classical music as accessible to any layperson imbued with the will and time to gain deep appreciation and appropriate understanding, whereas we assume that science must remain impenetrable, even to potentially interested people who could no more twirl a laboratory dial or manipulate a double integral than I could match Pavarotti (in his prime) performing Puccini? One need not practice at the highest level in order to understand in a quite sophisticated manner—both in music and in science. Yet we grant the accessibility of *Nessun dorma*, while denying similar status to $E = mc^2$.

I regard the arcaneness and inaccessibility of science as pure mythology, unfortunately abetted by some conventional aspects of scientific practice (but also countered by others, unfortunately not so visible, or not so readily acknowledged as part of science). I believe—and have attempted to put into practice in some fifteen books of general writing—that even the most complex and sophisticated scientific concepts can be explained in fully accessible layman's language without any dumbing down, or loss of the detail and technical concepts required for genuine understanding.

Moreover, I regard this genre of popular writing as an essential part of the humanistic tradition, and not as an exercise in discourteous and ultimately distorting gee-whiz simplification. This collective effort of the centuries, after all, includes some truly noble precedents that ought to give everyone hope for a generally successful outcome and enterprise—most notably Galileo's decision to write his two greatest works, including the Copernican document that hastened his political undoing, as Italian dialogues for ordinary readers, rather than as Latin treatises for university scholars and clerics. (Newton's *Principia*, on the other hand, does remain generally unreadable both for its Latin and its mathematics.) We must also honor Darwin's wise and fair decision to write the *Origin of Species* as an eminently readable book for the general public, not as a technical monograph for scientists.

The second issue of science perceived as ugly and immoral, rather than merely arcane and inaccessible, causes even deeper trouble, but actually enjoys a simpler resolution, at least in principle. I do not deny the cardinal observation that any major increase in technological capacity also breeds potential for potent misuse as the evil twin of intended benevolence. A medieval Hitler, armed only with a crossbow, just couldn't inflict as much damage, or at least not nearly so quickly, as his modern counterpart with a nuclear bomb or a hijacked airplane pressed into ghoulish service as an explosive guided missile. And I cannot deny that science always serves as the main spur for technological growth.

But we must also emphasize a common distinction that cannot be downplayed as overly fine or self-serving for science, but that represents a proper allocation of ultimate responsibility. (I present here an epitome for the crucial argument that I shall develop more fully in chapter 9.) Science, by its very nature as a quest for factual understanding and explanation, cannot prescribe a moral resolution to any question. All the tragedies falsely laid at the doorstep of science arise from our moral and political failures. I admit, of course, that science impacts our moral discourse in at least two crucial ways. First, several deep dilemmas of morality only existed in abstract form, or never entered our consciousness at all, before science provided the tools for emergence into practicality. One cannot, as an obvious example, advocate the moment of conception as the *ethical* definition of life's beginning (there can be no unambiguous *factual* "beginning" of life in such an unbreakable continuum of biological events) until one understands, and can identify, the biology of conception. In fact, and in the absence of this knowledge, legal and moral authorities, during most of Christian history, accepted the quickening (or movement) of the foetus in the womb as the defining point (and first clear indication) of life's beginning—and abortion before this advanced moment in pregnancy did not then count as illegal or immoral by theological standards. But no study of the biology of conception and pregnancy can specify the ethical, theological, or merely political "moment" of life's legal or moral inception.

Second, the sheer efficacy of science forces our immediate attention, by greatly raising the stakes and the speed of potential destruction, to ethical and political issues that had not intruded themselves into the forefront of our consciousness (however much we understood them as abstractions or potential dangers for the future). In the most obvious example, we knew about anthropogenic extinction, by experience ranging from the death of the dodo in the late seventeenth century to the passenger pigeon at the beginning of the twentieth century. But, as a direct consequence of technology for clearing land and altering environments, the speed and extent of extinction have now quickened to a point where, without hype, we may proclaim our current residence within the sixth great mass extinction of life's geologic history. And since the motto of the environmental movement—extinction is forever—represents factual reality, not emotional hype, salvation truly becomes a question of now or never. (Species, as unique biological entities, built by evolution through millions of years, cannot be substituted or replaced like worn-out automobile tires. If we lose half the world's species, we will all be impoverished on a depauperate planet. A city that supplements its human inhabitants only with pigeons, rats, and cockroaches cannot succor our spirits or honor the magnificent diversity of evolution.)

But even though the impact of science forces attention to the ethical dimensions of these present dangers, we must firmly reject the common, yet utterly false, inference that science itself, by its very nature, must be irreligious, immoral, or inherently opposed to aesthetic urges and sensibilities. Science operates in the different domain of factual understanding. Any full human life (the hedgehog's one true way of wisdom) must be enriched by all these independent dimensions, and their fecund interactions: ethical, aesthetic, spiritual, and scientific (the fox's range of independent and necessary contributions).

Science, as mentioned several times before in this book, quickly reaches its logical limit when confronting these other magisteria of our full being. Science, for example, can proceed no further than the *anthropology of morals*. That is, we may document the relative frequencies, and the stated rationales, for various moral beliefs among our diverse cultures. We may even speculate about the evolutionary value of certain common practices in our original Darwinian status as hunters and gatherers on the African savannas. We do want to contemplate this information, if only to learn the limits of human flexibility, and to understand which moral decisions might be hard to institute, and which more easy. But science, in principle, can say nothing about the *morality of morals*. For even if we can show that a certain belief (in infanticide, or genocide under certain circumstances, for example) arose for Darwinian advantages under natural selection, and still remains acceptable to the majority of human cultures, these factual claims cannot, in any way, impart ethical validity to the behavior. We can only reject such practices by the strength of our moral reasoning. At best, the factual knowledge of science might help us to understand the difficulties we must face in our struggle to reach this proper ethical standard, and might even suggest some useful strategies for winning such general consent.

Similarly, our sincere acknowledgment that factual science cannot intrude upon spiritual questions about the meaning and value of life, as properly asked by theologians, frees us from enmity in two important ways. First, and logically, this separation of the factual from the spiritual allows a proper pursuit of appropriate expertise in each magisterium, without anger inspired by poaching, and with a prospect for effective dialogue based upon mutual respect.

Second, and practically, science can only lose in contemporary America if we falsely claim a decisive voice in ethical or theological debates. For reasons that I do not pretend to understand, America stands alone among Western nations in the testimony of an overwhelming majority of citizens that belief in a fairly conventional form of Supreme Being occupies a central position in their lives. (I confess that I see little sign of any practical impact for such a conviction, as expressed in any superior moral consciousness or seriousness of purpose in commitment to helping one's fellows. But I do not doubt the sincerity of the stated conviction for an instant. If people insist that such a belief occupies a central position in their lives, then, by God, it does.) Given this firm sociological fact, if religious people then come to believe that science stands in intrinsic opposition to their spiritual convictions, then, if I may lapse into the vernacular, science is screwed. Our best strategy—and the intellectually soundest and most honest position in any case, by the first argument—therefore requires genuine respect for these religious convictions (which a high percentage of scientists also share), and continual insistence that science cannot pose any threat to these central pillars of life's emotional support.

In short, and to express the "sound bites" of my three arguments, science needs the humanities to teach us the quirky and richly subjective side of our own enterprise, to instruct us in optimal skills for communication, and to place proper boundaries upon our competencies—so that we can all work together, for the best of humanity, uniting our factual skills with our ethical wisdom to form a shield and weapon in this age of immediate danger.

7

Sweetness and Light
as Tough and Healing Truth

TO CLOSE THIS PART OF THE BOOK WITH A STORY FROM ITS BEGINNING,
I wish to return to the late-seventeenth- and early-eighteenth-century debate
between Ancients and Moderns, and give a last word to the "other" side. I discussed the best argument and healing hand of the scientific Moderns in presenting Bacon's paradox about the old age (and consequent wisdom) of our
present; and Newton's aphorism, admitting the puny status of an infant called
science by arguing that we can now see farther only because we stand upon
the shoulders of Ancient giants. But I cannot cash out my argument for mending the ancient breach between science and the humanities both by stressing
the commonalities and by merging the different strengths of both sides, unless
I also give a fair hearing to the best polemic from the Ancient side—if only
to show that the cogency of a good case can prevail even in pugnacity, and
that even such a vigorous defense leaves abundant space for the joining and
healing here proposed.

My choice of a closing tale also emerges from another motive, at the same
time both highly specific and entirely general. The phrase "sweetness and

light" resides in my earliest memories, because my mother loved the image, and frequently cited the verbal conjunction. But I confess that, while approving the sentiments, I always viewed the epigram as wimpish and rather meaningless, however warm the feelings so invoked for purely personal reasons. What, after all, could be more vague and less than a favoring of something so obviously virtuous as good taste and bright vision?

But then, as an adult pondering the issue of disciplinary divisions, I discovered the source of this apparently innocuous and universal phrase. (I hadn't even supposed, I confess, that the phrase had a specific origin—for something so evidently virtuous need only "be," and need not claim a specific point of invention.) But then I discovered that "sweetness and light" not only boasted an interesting inception, but had also been devised as a motto to represent something quite partisan and specific, not merely to express a bland and obvious eternal verity. For the phrase explicitly cited the best human uses of two substances manufactured by bees—honey and wax, yielding sweetness (still in vigorous supply) and light (at least before Mr. Edison). And the bee responsible for the phrase emerged from the decidedly bitter pen of a master satirist and champion of the Ancients: Jonathan Swift, who invented this particular creature as a metaphor to carry the case of the Ancients against the Moderns, the latter epitomized in the same fable by a spider. So sweetness and light summarizes the brief of classical humanism against the new world of science in its pugnacious infancy. And if Mr. Swift's posthumous (and undoubtedly still pugnacious) self will excuse the metaphysical expropriation, I would also like to apply his famous phrase to describe the *summum bonum* that would arise from the careful (see the last line of this book's preface) and *achievable* joining, in respectful independence, of the sciences and humanities by using the different (and equally excellent) stratagems of the fox and hedgehog.

We no longer take sides, but must find a way to mediate and merge these two great and truthful ways. In other words, we must generate sweetness and light from the totality of the bee and the spider, and not hold this prize as a weapon of one side against the other. But to grasp the power of this expansion, we must first know the true story of sweetness and light,* presented in

---

*The remainder of this chapter is an edited version of my previous essay, inevitably titled "Sweetness and Light," and published in my book *Dinosaur in a Haystack* (New York, Harmony Books, 1995).

the fundamental source, Jonathan Swift's celebrated satire of 1704, "A Full and True Account of the Battle Fought Last Friday Between the Ancient and the Modern Books in St. James's Library," usually called, for short, the "Battle of the Books." All might have been well had the two parties made a concordat, and kept to their own proper spaces. But the librarian had fostered discord by intemperate mixtures in shelving: "In replacing his books, he was apt to mistake, and clap Descartes next to Aristotle; poor Plato had got Hobbes . . . and Virgil was hemmed with Dryden."

Early in the text, both sides use Bacon's paradox to advance their respective arguments:

> Discord grew extremely high, hot words passed on both sides, and ill blood was plentifully bred. Here a solitary ancient, squeezed up against a whole shelf of moderns, offered fairly to dispute the case, and to prove by manifest reasons, that the priority was due to them, from long possession. . . . But these [the moderns] denied the premises, and seemed very much to wonder, how the ancients could pretend to insist upon their antiquity, when it was so plain (if they went to that) that the moderns were much more the ancient of the two.

The bulk of Swift's text describes the actual battle, with his own sympathies for the Ancients scarcely hidden—as in this passage, where Aristotle misses Bacon and kills Descartes instead (as the greatest French Modern falls into a vortex of his own theory):

> Then Aristotle, observing Bacon advance with a furious mien, drew his bow to the head, and let fly his arrow, which missed the valiant modern, and went hizzing over his head; but Descartes it hit. . . . The torture of pain, whirled the valiant bowman round, til death, like a star of superior influence, drew him into his own vortex.

Swift introduces the actual battle with a verbal curtain-raiser—a three-page gem that forms one of the greatest extended metaphors in Western literature: the dispute of the spider (representing the Moderns) and the bee (the Ancients). In the library, a spider dwells "upon the highest corner of a large window." He is fat and satisfied, "swollen up to the first magnitude, by the destruction of infinite numbers of flies, whose spoils lay scattered before the gates of his palace, like human bones before the cave of some giant." (Swift, I assume, did not know that males of most orb-weaving spiders are small and do not build webs—and that his protagonist was undoubtedly a "she." So, for that matter, come to think of it, is the industrious bee, also called "he" in this text.)

Swift clearly identifies the allegiances of his protagonists. The spider, spinning such a mathematically sophisticated web from his own innards (not relying on any external source of succor), is a scientific Modern:

> The avenues to his castle were guarded with turnpikes, and palissadoes, after all the *modern* way of fortification [Swift's own italics]. After you had passed several courts, you came to the center, wherein you might behold the constable himself in his own lodgings, which had windows fronting to each avenue, and ports to sally out upon all occasions of prey and defense. In this mansion, he had for some time dwelt in peace and plenty.

A bee then flies through a broken pane and happens "to alight upon one of the outward walls of the spider's citadel." His weight breaks the spider's web, and the convulsions of the resulting tumult awaken the spider, causing him to run out in fear "that Beelzebub with all his legions, was coming to revenge the death of many thousands of his subjects, whom the enemy had slain and devoured." (A nice touch. Beelzebub, a popular name for the devil, is literally "lord of the flies.") Instead he finds only the bee, and curses in a style that has been called Swiftian ever since: "A plague split you . . . giddy son of a whore. . . . Could you not look before you, and be damned? Do you think I have nothing else to do (in the Devil's name) but to mend and repair after your arse?"

The spider, calming down, now takes up his intellectual role as a Modern and excoriates the bee with the crucial argument from his side: You advocates of the Ancients operate as pitiful and unoriginal drones who create nothing yourselves, but can only forage among other people's antique insights (the flowers in the field, including nettles as well as objects of admitted beauty). We Moderns build new intellectual structures from the heart of our own genius and discovery:

What art thou, but a vagabond without house or home, without stock or inheritance? Born to no possessions of your own, but a pair of wings, and a drone-pipe. Your livelihood is an universal plunder upon nature; a freebooter over fields and gardens; and for the sake of stealing, will rob the nettle as readily as a violet. Whereas I am a domestic animal, furnished with a native stock within myself. This large castle (to show my improvements in the mathematics) is all built with my own hands, and the materials extracted altogether out of my own person.

The bee then responds for all devotees of ancient learning: I borrow, but cause no harm in so doing, and I transmute what I borrow into new objects of great beauty and utility—honey and wax. But you, while claiming to build only from your own innards, must still destroy a hecatomb of flies for the raw material. Moreover, your vaunted web is weak, temporary and ephemeral, whatever its supposed mathematical beauty (while the distillation of ancient knowledge endures forever). Finally, how can you claim virtue for a product of your own spinning if the material be poison based on your own gall, and the effect thereof destruction?

I visit, indeed, all the flowers and blossoms of the field and the garden, but whatever I collect from thence, enriches myself, without the least injury to their beauty, their smell, or their taste. . . .

You boast, indeed, of being obliged to no other creature, but of drawing and spinning out all from yourself; that is to say, if we may judge of the liquor in the vessel by what issues out, you possess a good plentiful store of dirt and poison in your breast; and, tho' I would by no means, lessen or disparage your genuine stock of either, yet, I doubt, you are somewhat obliged for an increase in both, to a little foreign assistance. . . . In short, the question comes to this; whether is the nobler being of the two, that which by a lazy contemplation of four inches round; by an overwhelming pride, which feeding and engendering on itself, turns all into excrement and venom; produces nothing at last, but fly-bane and a cobweb: or that, which, by an universal range, with long search, much study, true judgment, and distinction of things, brings home honey and wax.

No one has ever set forth the issue more incisively, albeit in extreme form, in nearly three hundred years of subsequent writing. Most thoughtful people come down somewhere between the bee and the spider, but extremists on both sides still invoke the same arguments. Current partisans of the spider claim that the "great books" of traditional learning (now including such former Moderns as Swift and his *Gulliver's Travels*) have become both unreadable and irrelevant for modern students—and might as well be dropped (or lightly retained as a few excerpts for a lick and a promise) in favor of direct engagement with modern literature and science. At worst, they may actively disparage the old mainstays as nothing but repositories of prejudice written by that biased subset of humans called dead white European males (or DWEMs for short).

Current partisans of the bee can dispense worthy platitudes about upholding standards and retaining a canon universally validated by endurance through so much time and turmoil. But these good arguments are often accompanied by blindness, or actual aversion, to the scientific and political complexities that permeate our daily lives and that all educated people must understand in order to be effective and thoughtful in their professions. Moreover, defense of the "great books" too often becomes a smokescreen for political conservatism and maintenance of old privileges (particularly among folks like me—white professors past sixty who don't wish to concede that other kinds of people might have something important, beautiful, or enduring to say).

How can we resolve this ancient debate from the youth of our modern time? In one sense we can't, at least to anyone's clear victory—for both sides present good arguments, following Bacon's paradox that once epitomized the ongoing struggle. But an obvious solution stares us all in the face, if only we could overcome the narrowness and parochiality that leads any partisan to fortify his barricade. The answer has been with us since Aristotle—in the form of the "golden mean." The solution speaks to us by compelling attention to good points on both sides. This answer lies embodied in the famous epigram of Edmund Burke (1729–1797), once a Modern in the original battle, but now an archconservative among the DWEMs: "All government—indeed, every human benefit and enjoyment, every virtue and every prudent act—is founded on compromise and barter." We must hybridize the bee and the spider—and then, in good Darwinian fashion, select for the best traits of both parents in a rigorous program of good breeding (education). The spider surely prevails in extolling the technical beauty of his web, and the absolute need for all contemporary people to understand both the mechanics and aesthetics of its structure. But the bee cannot be faulted for insisting that fields of well-distilled wisdom await our entirely benign exploitation for enjoyment and enlightenment—and that we would be utter fools to bypass such a rich store-house.

I can argue the virtues of both sides, but since I live in the world of science, and experience its parochialities on a more sustained and daily basis, I feel more impelled to advance the bee's cause. Distillation may be biased, but anything that endures for hundreds or thousands of years (at least in part by voluntary enjoyment rather than forced study) must contain something of value. No one celebrates diversity more than evolutionary biologists like myself; we love every one of those million beetle species, every variation in every scale on a butterfly's wing, every nuance in the coloration of each feather on a peacock. But without some common mooring, we cannot talk to each other. And if we cannot talk, we cannot bargain, compromise, and understand. I am sad that I can no longer cite the most common lines from Shakespeare or the Bible in class, and hold any hope for majority recognition. I am troubled that the primary lingua franca of shared culture may now be rock music of the last decade—not because I regard the genre as inherently unworthy, but because I know that the language will soon change and therefore sow more barriers to intelligibility across generations. I am worried that people with inadequate knowledge of the history and literature of their culture will ultimately become entirely self-referential, like science fiction's most telling symbol (from E. A. Abbott's *Flatland,* published in 1884 and in print ever since)—the happy fool who lives in the one-dimensional world of point-land, and thinks he knows everything because he forms his own entire universe. In this sense, the bee criticizes the spider properly—an ephemeral cobweb "four inches round" can only provide a paltry sample of our big and beautiful world. I can't do much with a student who doesn't know multivariate statistics and the logic of natural selection; but I cannot make a good scientist—though I can forge an adequate technocrat—from a person who never reads beyond the professional journals of his own field. Any genuinely wise person will have to know and appreciate the truly different ways of the sciences and humanities in order to achieve an *integral* excellence. Bee plus spider; the fox's way to become an optimal hedgehog. Difficult—but surely possible in our new age of genetic engineering!

I give the last word to Swift. When the bee and the spider finish their argument, Aesop steps up and praises both parties, who have "admirably managed the dispute between them, have taken in the full strength of all that is to be said on both sides, and exhausted the substance of every argument pro and con." But he then, as befits his station and status, supports the bee. A person who ignores accumulated wisdom perishes in his own thin web:

> Erect your schemes with as much method and skill as you please; yet if the materials be nothing but dirt, spun out of your own entrails (the guts of modern brains) the edifice will conclude at last in a cobweb: the duration of which, like that of other spider webs, may be imputed to their being forgotten, or neglected, or hid in a corner.

Aesop ends by praising the bee and inventing a proverb based upon one of the loveliest conjunctions in English. And thus did the phrase "sweetness and light"—as direct properties of honey and wax—enter our lexicon of sayings as the culmination of Swift's defense, via Aesop, for the extended hive of our greatest intellectual traditions.

> As for us, the Ancients; we are content with the bee, to pretend to nothing of our own, beyond our wings and our voice: that is to say, our flights and our language; for the rest, whatever we have got, has been by infinite labor, the search and ranging through every corner of nature: the difference is, that instead of dirt and poison, we have rather chose to fill our hives with honey and wax, thus furnishing mankind with the two noblest of things, which are sweetness and light.

III

A SAGA OF *PLURIBUS* AND *UNUM*

The Power and Meaning
of True Consilience

8

The Fusions of *Unum* and the Benefits of *Pluribus*

MY BRIEF FOR THE OBLITERATION OF HARMFUL BOUNDARIES AND mutual suspicion between science and the humanities includes two recommendations that may seem contradictory at first—but no more so than the official motto of our nation: *E pluribus unum* (one from many). We fought a civil war to keep our diverse themes together, to prove that one nation, strong and democratic, could include a full range of human and natural differences— ethnic, linguistic, climatic, economic, topographical—under a single canopy of mutual respect. So too for our disciplinary domains in a united realm of the human intellect, and especially for the perceived conflict of science and the humanities. We can break these old bonds of recrimination, and become equal partners in unity, if we practice, simultaneously, both sides of a superficial contradiction with a deeper underlying consonance: that is, if we can enjoy our fusion in intentions, motives, and several aspects of creative practice (the hedgehog's one great way), but also respect our discreteness and separation as guardians of distinct magisteria charged with the exploration of logically different kinds of questions (the fox's many effective but separate ways).

Two quotations about diversity, one from within and one from without, summarize the case for mutual respect with acknowledgment of defining differences and also a set of likenesses rooted in the commonalities of all intellectual effort. First, and from within, each of the domains or magisteria embodies, inside its own being, so many different methods, concerns, and styles of explanation that no knee-jerk united front could be contrived even if we wanted to wage war under a monistic banner. (This book treats science and the humanities, but the same argument applies to other domains, notably religion.) Each magisterium embraces its own *E pluribus unum,* and each can only be harmed by struggles for supremacy from within. How, then, could the entire collectivity hope to profit by the same kind of destructive struggle with other distinct collectivities? The anthropologist Clifford Geertz emphasized this practical power of pluralism in a commentary for *Science* (July 6, 2001, page 53), the leading American journal for professionals in the trade. Interestingly, Geertz invokes the phony "science wars" (discussed herein on pages 95–104) to introduce his important observation about widespread diversity within magisteria:

> For the most part, "the science wars," trafficking in tribal jealousies and archaic fears, have produced more heat than light. But in one respect they have been useful. They have made it clear that using the term "science" to cover everything from string theory to psychoanalysis is not a happy idea, because doing so elides the difficult fact that the ways in which we try to understand and deal with the physical world and those in which we try to understand and deal with the social one are not altogether the same. The methods of research, the aims of inquiry, and the standards of judgment all differ, and nothing but confusion, scorn, and accusation—relativism! Platonism! reductionism! verbalism!—results from failing to see this.

Second, and from without, I have long appreciated the wise and, at first glance, paradoxical observation of G. K. Chesterton about art, but equally applicable to the definition of any legitimate discipline. For, in the absence of well-defined boundaries, no organism or institution can maintain sufficient coherence for recognition as a legitimate entity at all. Chesterton (1874–1936), now remembered primarily for his Father Brown series of mystery stories, was a respected essayist and perhaps the most famous literary critic of his time. He wrote: "Art is limitation; the essence of every picture is the frame."

In keeping with my practice throughout this book, I will forgo any further abstract or theoretical discussion in favor of particular examples, not widely known, that strike me as especially apropos or poignant in illustrating the general thesis under discussion. Thus I will pursue my two apparently contradictory, but actually complementary, themes for union and cooperation between the sciences and the humanities (the fusions of *unum* and the benefits of *pluribus*) by presenting two examples within each category.

THE FUSIONS OF *UNUM*

HAECKEL'S "ARTFORMS OF NATURE"— EITHER OR NEITHER? FUSED OR MISUSED?

The power of many important works in the history of Western art and science has been greatly enhanced by a fusion so intimate that inquiries into whether the product should be called "art" or "science" cease to make any sense at all— for "neither" or "both" provide equally cogent answers, thus proving that the question itself has become meaningless because the two putative categories of this false dichotomy do not exist as separate and competing entities.

In my favorite example of maximal fusion, the German biologist Ernst Haeckel (1834–1919) also worked as a quite competent painter and graphic artist. (Of course, many scientists have tried their amateur hands at art, but only as "Sunday painters" in the usual dismissive description. Goethe, for example, produced large numbers of eminently forgettable watercolors. But at least two celebrated naturalists also possessed the fortunate gift of genuine artistic skill, and their technical publications, featuring their own illustrations, gain great strength through the conjunction—Ernst Haeckel and the great French naturalist Georges Cuvier.)

In 1904, Haeckel published a magnificent volume of exactly one hundred plates, titled *Kunstformen der Natur,* or Artforms of Nature. The title itself explicitly states an intention to treat the two great domains together. But the content of the plates realizes this goal to a degree never before attained in the history of scientific illustration—unleashing the paradox that such a superb realization also extinguished the category of "scientific" illustration, thus so outstandingly treated! From 1899 to 1904, as Haeckel produced his plates in ten installments of ten each, Art Nouveau, called *Jugendstil* in Germany, reigned as the height of fashion in the fine and decorative arts. In rough epitome, the *Encyclopaedia Britannica* states that "the distinguishing ornamental characteristic of Art Nouveau is its undulating, asymmetrical line, often taking the form of flower stalks and buds, vine tendrils, insect wings, and other delicate and sinuous natural objects; the line may be elegant and graceful or infused with a powerfully rhythmic and whiplike force."

If we approach the plates of *Kunstformen der Natur* with the conventional question—is this art or science?—we scarcely know how to respond. Haeckel does depict real creatures that actually exist, so the plates, in one sense, promote science. But both the individual organisms themselves, and their layout on each plate, rigorously obey all the key conventions of Art Nouveau, with sinuously extended curves everywhere—so the plates, in another sense, embody the prevailing artistic style of the time.

Consider just three examples (I would love to reproduce all one hundred plates—and in color; but my publisher would demur, and the work does remain in print, in a mediocre reproduction of the plates by Dover Books). The squid and octopuses of figure 24 do exist, and we know that these creatures grow long and numerous tentacles. But I doubt that any of their natural poses include such conformity to the preferred swirls of Art Nouveau. For the glass sponges of figure 25, Haeckel does show the angular symmetry of the nearly microscopic spicules building the internal skeleton of silica. But the mixed tableau of several species in their entirety (at the bottom of the plate) might have been commissioned by an art teacher as the instruction manual for a favored style then honored as the height of fashion. And when Haeckel doesn't merely gang together a group of individual organisms, but attempts to construct a "natural" scene of numerous species in their habitats (as in figure 26 of reef-building corals), the ensemble looks more like a unified phantasmagoria of Art Nouveau curvatures than an array of separate and living organisms.

I find Haeckel's spare commentary, presented as introductory and closing statements about the plates, particularly revealing in expressing both his satisfactions and his disquiet. He clearly states, as the fusion of his title proclaims, that he wished to unite both artistic and scientific goals in a single series of illustrations (my translations from his German):

Figure 24.

The primary purpose of my *Artforms of Nature* was aesthetic: I wanted to provide an entry, for a wider circle of people, into the wonderful treasures of natural beauty hidden in the depths of the sea, or only visible, as a consequence of small size, under the microscope. But I also wanted to combine these aesthetic concerns with a scientific goal: to open up a deeper insight into the wonderful architecture of the unfamiliar organization of these forms.

But Haeckel could not rest content in this love-fest of fusion—for he knew that he faced a problem with his scientific colleagues (Haeckel's primary

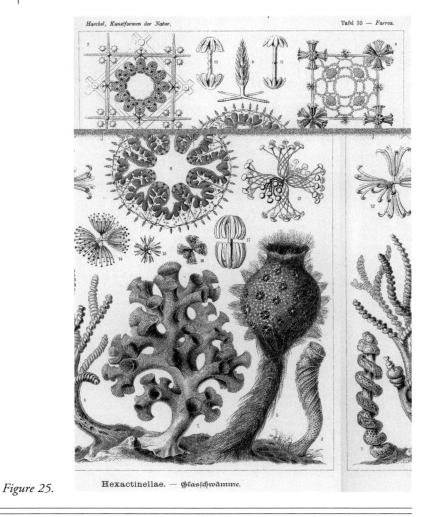

Figure 25.   Hexactinellae. — Glasſchwämme.

day job, after all) who would surely pounce in derision upon any distortion of biological accuracy, presumably, indeed especially, for art's sake. In fairness, one cannot accuse Haeckel's colleagues of narrow parochialism in their strict scrutiny of his work. For decades Haeckel had been justly criticized for his cavalier attitude toward accuracy, even in his technical publications for taxonomic specialists. In particular, he often "improved" the geometric symmetries of radiolarian skeletons and sponge spicules, confecting forms of unerring regularity and beauty to replace actual creatures only slightly less attractive in their not-quite-perfect symmetry. More important, Haeckel had been roundly

Figure 26.

and rightly rebuffed for his frequent practice of drawing idealizations for his textbooks, and claiming them as actual specimens. In the most notorious example, exposed almost immediately by several colleagues (see my article on Louis Agassiz's reaction in *I Have Landed,* Harmony Books, 2002), Haeckel supported his favorite hobbyhorse—the so-called "biogenetic law" of "ontogeny recapitulates phylogeny"—by simply drawing the same figure three times as a supposed illustration of the near identity in early embryonic form among vertebrate species of highly disparate adult design! (Following the old dictum about persistence of bad pennies, modern creationists have reexhumed this more-than-twice-told, and well-castigated, tale in a rearguard action to cast doubt on evolution because a distinguished colleague misbehaved in this manner more than a century ago.)

I would, however, criticize Haeckel's colleagues for cynically invoking the parochial ploy that Haeckel's flouting of scientific norms flowed from an "artistic" bent, thereby distorting his commitment to scientific accuracy. Why should an artist be any less concerned about veracity than a scientist? Such stereotypes, unfortunately all too familiar and persistent in our times as well, can only poison the pluralism and respect sought by all people of goodwill. Haeckel's failures lie at the doorstep of his own inadequacies, and cannot be fobbed off on the general practice of any larger group counting him among its membership.

Haeckel therefore exposed his own fears in a gender-bending version of "the lady doth protest too much, methinks," defending himself vociferously, on an issue that might otherwise have passed in silence, by insisting that all his plates depicted actual animals in detailed accuracy. Yet, in the *Kunstformen*, Haeckel had, more than ever before (but justifiably for once, given the intention of the work), consistently distorted his organisms by arranging their parts in unnatural swirls, and by conflating creatures into impossible conjunctions based on felicity of design—all, and obviously, to match the reigning sensibilities of Art Nouveau, and not because such scenes ever existed in nature.

Perhaps I am reading too much into minor stylistic matters, but the differences between Haeckel's two statements in his own defense—one in 1899 at the outset of his work, and the other in 1904 at the completion—seem to reveal an increasing need to secure his colleagues' understanding by toeing the expected scientific line. In 1899 he wrote in the active voice of ordinary prose, using the dreaded first person singular, usually shunned in scientific texts, and clearly leaving some space for departure from factuality to the discretion of "real" artists:

> In these figures, I have restricted myself to objects of nature that truly exist, and I have refrained from all stylistic modeling and decorative uses; I leave these devices to the artists themselves.

But in 1904, as if to distance himself from his own productions, and now bowing to conventions of scientific prose, Haeckel makes the same point in the passive voice, with no dispensation awarded to artists for any departure from nature's truth: "All the 'Artforms' depicted here are, in truth, forms that really exist in nature; and they have been drawn without any idealization or stylistic license."

NABOKOV'S BUTTERFLIES: CLARITY IN FACT

If this first example of the fusions of *unum* cites a case so intermixed and inter-mediate that the conventional labels of "art" and "science" lose all meaning as distinct modes of inquiry, then a second form of fusion, less intense but far more common, uses the ordinary skills and sensibilities of the "other" side to enhance effective argument in a "home" domain of conventional expertise (often beyond the explicit notice of more parochial practitioners). I have already discussed how such preeminent figures as Charles Lyell and Sigmund Freud advanced their causes by employing an uncommon gift for writing powerful and stylish prose—a "tactic" that many scientists would regard as "stealth," or at least as irrelevant to conventional standards of rigor in data and logic of argument. (Needless to say, I do not claim either that all scholars in the humanities write skillfully, or that scientists don't favor well-wrought over disorganized prose. I only point out that humanists explicitly value good writ-ing as a primary desideratum of their enterprise, whereas most scientists tend to dismiss stylistic matters as essentially irrelevant to their work.)

My favorite example in this second category of *unum* cites the fascinating case of a great literary figure of the twentieth century (and also a more than merely competent biologist) who followed an important norm of science in his literary work, in full knowledge of what he did, why he so proceeded, and how his writing would be enhanced thereby. Nonetheless, nearly all literary critics have failed to understand either the strategy or the reasons (even though the author stated his aims, explicitly and often), and have maintained their stubborn allegiance to a conventional "literary" explanation that the author himself loathed and rejected. An ironic tale indeed, well fit for the full range of lessons, from moral to political.

Vladimir Nabokov* worked from 1942 to 1948 as curator of lepidoptery (butterflies and moths) in the Museum of Comparative Zoology at Harvard University, three floors above the office that I have occupied in the same build-

---

*The closing tales of this section include some material from previously published essays: Nabokov from *I Have Landed* (Harmony Books, 2002), Thayer from *Bully for Brontosaurus* (W. W. Norton, 1991), and Poe from *Dinosaur in a Haystack* (Harmony Books, 1995).

ing for thirty-five years. He was a skilled and fully professional specialist on the taxonomy and natural history of the Polyommatini, popularly known as "blues," and he published several respected technical monographs on this large group of Latin American butterflies. In fact, as his biographers often remark, before 1948, when he began to teach literature at Cornell, Nabokov earned his primary living, and spent most of his time, as a biologist—and would justly have been labeled a professional scientist and amateur author.

We can scarcely doubt Nabokov's love for his first profession, as eloquently expressed in a 1945 letter to his sister:

> My laboratory occupies half of the fourth floor. Most of it is taken up by rows of cabinets, containing sliding cases of butterflies. I am custodian of these absolutely fabulous collections. We have butterflies from all over the world. . . . Along the windows extend tables holding my microscopes, test tubes, acids, papers, pins, etc. I have an assistant, whose main task is spreading specimens sent by collectors. I work on my personal research . . . a study of the classification of American "blues" based on the structure of their genitalia (minuscule sculpturesque hooks, teeth, spurs, etc., visible only under the microscope), which I sketch in with the aid of various marvelous devices, variants of the magic lantern. . . . My work enraptures but utterly exhausts me. . . . To know that no one before you has seen an organ you are examining, to trace relationships that have occurred to no one before, to immerse yourself in the wondrous crystalline world of the microscope, where silence reigns, circumscribed by its own horizon, a blindingly white arena—all this is so enticing that I cannot describe it.

Following the fate of many scientists who spent years in ceaseless scrutiny and drawing of delicate anatomical features under the microscope, Nabokov's vision became so impaired that he could no longer pursue the detailed work he loved. Yet, and poignantly, he stated in a 1975 interview, long after he had ceased his biological research, that the lure and passion remained as strong as ever:

> Since my years at the Museum of Comparative Zoology in Harvard, I have not touched a microscope, knowing that if I did,

I would drown again in its bright well. Thus I have not, and probably never shall, accomplish the greater part of the entrancing research work I had imagined in my young mirages.

Because Nabokov ranks among the aesthetic gods of our time, critics and scholars have sifted every word of his writing for clues about sources and influences, and a veritable "industry" of Nabokovian interpretation has constructed elaborate and implausible literary "theories" about the meaning of his work. In reading through this material for an essay on Nabokov's lepidoptery in literature (published in *I Have Landed,* Harmony Books, 2002), I became both amused and a bit disturbed by the inability of most literary scholars to think outside their own "box" and to proceed beyond their conventional modes of interpretation. All critics recognize, of course, that Nabokov's writing includes copious references to butterflies and moths, and all scholars know the sources of Nabokov's expertise in this biological arena.

Faced with a consequent need to examine the relationship between Nabokov's science and his writing, scholars in the humanities have, almost invariably, taken refuge in the conventional claim of their craft, despite Nabokov's own clear rejection of this hypothesis. They argue that, as a literary man, Nabokov used his knowledge of butterflies primarily as a source for metaphors and symbols. Joann Karges, for example (in *Nabokov's Lepidoptera: Genres and Genera,* Ardis Press, 1985) writes: "Many of Nabokov's butterflies, particularly pale and white ones, carry the traditional ageless symbol of the anima, psyche or soul . . . and suggest the evanescence of a spirit departed or departing from the body."

But Nabokov himself vehemently insisted that he not only maintained no interest in butterflies as literary symbols, but also that he would regard such usage as a perversion and desecration of his true concerns. (Artists, and all of us, of course, have been known to dissemble, but I see no reason to suspect Nabokov's explicit comments on this subject.) For example, he stated in an interview: "That in some cases the butterfly symbolizes something (e.g., *Psyche*) lies utterly outside my area of interest."

Over and over again, Nabokov debunks symbolic readings in the name of respect for factual accuracy. For example, he criticizes Poe's metaphorical invocation of the death's-head moth because Poe didn't describe the animal and, even worse, because he placed the species outside its true geographic range: "Not only did he [Poe] not visualize the death's-head moth, but he was also under the completely erroneous impression that it occurs in America." Most tellingly, in a typical Nabokovian passage in *Ada,* he playfully excoriates Hieronymus Bosch for including a butterfly as a symbol in his *Garden of Earthly Delights,* but then depicting the wings in reverse by painting the gaudy top surface on an insect whose folded wings should be displaying the underside!

> A tortoiseshell in the middle panel, placed there as if settled on a flower—mark the "as if," for here we have an example of exact knowledge of the two admirable girls, because they say that actually the *wrong* side of the bug is shown, it should have been the underside, if seen, as it is, in profile, but Bosch evidently found a wing or two in the corner cobweb of his casement and showed the prettier upper surface in depicting his incorrectly folded insect. I mean I don't give a hoot for the esoteric meaning, for the myth behind the moth, for the masterpiece-baiter who makes Bosch express some bosh of his time, I'm allergic to allegory.

Finally, when Nabokov does cite a butterfly in the midst of a metaphor, he attributes no symbolic meaning to the insect, but only describes an accurate fact to carry his more general image. For example, he writes in an early story, titled "Mary": "Their letters managed to pass across the terrible Russia of that time—like a cabbage white butterfly flying over the trenches."

I think that we should accept Nabokov at his own word, and honor his different interpretation of how his scientific sensibilities played out within his literature—or rather, and more accurately, how a crucial aspect of his temperament, and a central component of his convictions, served him so well, and in the same manner, in both his fiction and his science. Nabokov, as one of literature's consummate craftsmen, upheld the sacredness of accurate factuality—an obvious requirement in science, but also a boon to certain genres of literature. Interestingly, and befitting his deservedly greater reputation as a writer than as a biologist (for Nabokov ranks as one of the great novelists of all time, and as an accomplished technician, but not as a brilliant theorist, in science), Nabokov frequently asserted—thus placing his story within this section on the fusions of *unum*—that literature and science meet in mutual respect for detailed factuality, with the highest virtue of accuracy residing in the evident beauty of such material truth.

Thus no one grasped the extent of underlying unity between science and literature better than Vladimir Nabokov, who worked with different excellences as a full professional in both domains. Nabokov often insisted that his literary and entomological pursuits shared a common mental and psychological ground. In *Ada,* while invoking a common anagram for "insect," one of Nabokov's characters states: " 'If I could write' mused Demon, 'I would describe, in too many words no doubt, how passionately, how incandescently, how incestuously—*c'est le mot*—art and science meet in an insect.' "

Returning to his central theme of aesthetic beauty in both the external existence and our internal knowledge of scientific detail, Nabokov wrote in 1959: "I cannot separate the aesthetic pleasure of seeing a butterfly and the scientific pleasure of knowing what it is." When Nabokov spoke of "the precision of poetry in taxonomic description"—no doubt with conscious intent to dissipate a paradox that leads most people to regard art and science as inexorably distinct and opposed—he used his literary skills in the service of unity. Thus in a 1966 interview Nabokov broke the boundaries of art and science by stating that the highest ideal of each domain must also characterize true excellence in the other:

> The tactile delights of precise delineation, the silent paradise of the *camera lucida,* and the precision of poetry in taxonomic description represent the artistic side of the thrill which accumulation of new knowledge, absolutely useless to the layman, gives its first begetter. . . . There is no science without fancy, and no art without facts.

THE BENEFITS OF *PLURIBUS*

The foregoing stories of Haeckel and Nabokov illustrate how foolishly we can waste our time, and how erroneously we may formulate our conclusions, when we fail to perceive the unified intent of creative acts, and insist upon categorizing them as either "art" or "science," under the fallacy that proper placement will clarify a true intention implicit in one field but actively abjured by the other (artistic license versus natural fidelity for Haeckel, or taxonomic factuality versus literary symbolism for Nabokov). In this section I will tell two stories of apparently opposite form, but of actually identical meaning. For, in both these cases, a persistent puzzle or an erroneous and uncomfortable interpretation (explicitly so identified) has pervaded our conventional literature about an important figure, because we have classified him into one of the two domains (the arts in both cases), whereas the simple solution to the long-standing error requires access to a little item of knowledge conventionally housed in the other domain (the sciences in both stories). In each case, the man himself operated as an *unum* who worked in both science and the humanities (and did not impose upon himself the dichotomy of "never the twain shall meet"); whereas the solution to the persistent scholarly puzzle requires that we bring together the *pluribus* of both his true concerns.

The two stories also present an interesting contrast in the opposite forms of their particular narratives. In the first tale—the valuable scientific work of the maligned artist Abbott Handerson Thayer—we learn about a man who solved a long-standing problem in natural history because, as an artist, the simple solution lay within his realm of learning and discourse, and simply hadn't been encountered by a professional naturalist. In the second tale—the disarmingly simple solution to an old conundrum about Edgar Allan Poe, including a novel affirmation of real value in his only scientific work—we learn that the solution to an issue that has bothered generations of literary scholars lies in a basic fact about the history of molluscan taxonomy, an item known to every practicing systematist of clams and snails, but never applied to Poe's problem because these scientists (who would have recognized the solution in a flash) had never encountered the problem (which had resided exclusively in the technical writings of literary critics). Thus, in the first case a visual artist uses his special tool to solve an old puzzle in science; in the second case, a particular fact of science solves an old puzzle about a literary artist.

THE LOWERED DIMENSIONALITY OF THAYER'S HIGHER INSIGHT

Abbott Handerson Thayer (1849–1921) does not rank as a household name these days, even among folks reasonably well versed in the history of American art. But, at the acme of his success, around the turn of the twentieth century, before the winds of modernism swept his ethereal paintings of angels and innocent children into oblivion, Thayer occupied the pinnacle of his profession. In fact, Thayer resides among the four contemporary artists (with James McNeill Whistler as the best remembered today), so favored by the industrialist Charles Lang Freer that this wealthy patron established his Freer Gallery, now a major museum within the Smithsonian Institution of Washington, D.C., specifically to house the work of this quartet along with his spectacular collection of oriental art. (Of course, worms turn and winds of fashion reverse. Thayer may never regain his former renown, but angels are certainly back in style these days, and the December 27, 1993, issue of *Time* magazine featured one of Thayer's best on its cover.)

Strangely enough, most evolutionary biologists also know something about Thayer, but in an entirely different context, and almost never by name. In my world, he has survived as a footnote of derision for a standard classroom lecture on the adaptive value of animal coloration, and on the dangers of carrying a pet theory too far. Thayer, who lived in rural New Hampshire, pursued his hobby of birdwatching with sufficient zeal and study to become a respected amateur naturalist and author of several technical articles in professional journals of ornithology. He also, as my story now begins to unfold, followed an all-too-common path in human conviction by developing a good idea about animal coloration, but then elevating his insight, first into a dominant theme, then into a pervasive phenomenon, and finally into an exclusive truth that tolerated not a single exception in the entire domain of nature.

Naturalists argue—no doubt correctly—that color patterns serve a variety of adaptive ends throughout the animal kingdom. In particular, many patterns conceal creatures from potential enemies, whereas other configurations of form and color serve the opposite function of announcing an animal's presence, perhaps to court a mate or to scare off other suitors. Thayer, in short, discovered several genuine examples of concealment, based on principles that had not been sufficiently recognized or understood by previous naturalists. Scientists generally gave warm credit to this initial work (mostly from the 1890s), sometimes with a touch of bemusement, or even begrudgement, because an artist had beat them at their own game—but still with praise and fair acknowledgment.

Unfortunately, Thayer then followed the Lorelei's beckoning song of an *idée fixe,* or one true way. He decided that, in principle and no matter how apparently to the contrary, *all* colors on the hides and hairs of animals, each and every last one, must have evolved for purposes of concealment, never to reveal or advertise. Thayer applied this exclusive principle throughout nature, from the ever-so-obvious stripes of the zebra (invisible, Thayer showed [see figure 27], in reeds where zebras do not in fact live, but so notable on the open plains, where they do reside), to the gaudy colors of the peacock's tail (which the bird, when courting, so evidently displays with panache to the peahen, whatever else he might do with the apparatus at other times). In any case, Thayer showcased his uncompromising and controversial views in an elaborate 1909 book, largely authored by his son Gerald: *Concealing-Coloration in the Animal Kingdom* (New York, Macmillan).

Thayer owes his continuing biological sound bite of derision to an argument that he himself acknowledged as his farthest stretch, but also recognized as necessary to imbue his theory with the full generality he so devoutly sought. Blotches and stripes, however prominent, could always be interpreted as efforts to conceal an animal by breaking its integrity into separate pieces (a common device of human camouflage). But Thayer recognized that monochromatic patterns, especially bright colors, posed special problems for interpretation as concealing devices. Hence, Thayer's conventional downfall in his gutsy, if improbable, attempt to explain the bright and monochromatic redness of flamingos. This color, Thayer argued in all seriousness, evolved to conceal the animals as they feed and blend into the ruddy colors of the rising or setting sun. And this gorgeously ridiculous application of a good theory, pushed maximally beyond its legitimate domain, has served generations of university lecturers as a paradigm for decent intentions unsuppressed by proper skepticism and application of the scientific method. But, to cite Thayer's own words:

> These birds are largely nocturnal, so that the only sky bright enough to show any color upon them is the more or less rosy and golden one that surrounds them from sunset till dark and from dawn until soon after sunrise. They commonly feed in immense, open lagoons, wading in vast phalanxes, while the entire real sky above them and its reflected duplicate below them constitute either one vast hollow sphere of gold, rose, and salmon, or at least glow, on one side or the other, with these tones. Their whole plumage is a most exquisite duplicate of these scenes. . . . This flamingo, having at his feeding time so nearly only sunrise colors to match, wears, as he does, a wonderful imitation of them.

FIG. 90.   Cardboard 'zebra' without stripes, against light straws as in Fig. 91.
Photograph, retouched.

FIG. 92.   Cardboard zebra among imitation reeds 'relieving' dark, as against the sky.
Photograph, retouched.

Figure 27.

Critics retaliated right away and to every detail. Flamingos do not concentrate their feeding at dawn and dusk, but remain active all day. Anacondas and alligators, their major enemies, do not inhabit the films of saline ponds that flamingos favor, and where Thayer thought they melted into invisibility at key times. Flamingos eat by filtering tiny eyeless animals, so the argument can't even operate in reverse to hide predatory flamingos from unsuspecting prey.

Far more generally (and embarrassingly), Thayer's argument must also fail on its own terms—and Thayer, who was overenthusiastic to a fault, but neither dishonest nor dishonorable, had to confess. Any object viewed *against* the fading light will appear dark, whatever its actual color. Thayer addressed this problem explicitly by painting a dark palm tree against the sunset in his infamous and fanciful painting of fading flamingos (reproduced here as figure 28, and for unfortunate practical reasons in inappropriate black and white). Thus he could only claim that flamingos looked like the sunset in the *opposite* side of the sky: red clouds of sunset in the west, red masses of flamingos in the east. Would any animal be so confused by two "sunsets," with flamingos showing dark against the real McCoy? Thayer admitted in his 1909 book:

> Of course a flamingo seen against dawn or evening sky would look dark, like the palm tree in the lower left-hand figure, no matter what his colors were. The . . . right-hand figures, then, represent the lighted sides of flamingos at morning or evening, and show how closely these tend to reproduce the sky of this time of day; although always, of course, <u>in the opposite quarter of the heavens</u> [Thayer at least fairly underscores his own admission] from the sunset or dawn itself.

Two other, and more specific, reasons beyond the almost perfect ridiculousness of Thayer's flamingo blunder has kept his story in active circulation. First, an old maxim for endurance (not Thayer's aim for this particular error!) cites the virtue of attracting famous adversaries—and Thayer could not possibly have surpassed himself on this score. Teddy Roosevelt (whom I once regarded, in my arrogant and ignorant youth, as an impostor on Mount Rushmore in the presence of Lincoln, Jefferson, and Washington, but whom I now regard as one of the most fascinating characters in American history) operated as a distinguished natural historian and avid big-game hunter, when not engaged in more mundane pursuits. Roosevelt also took a strong interest,

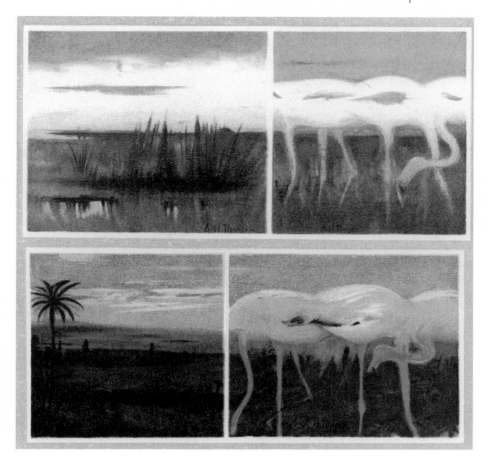

Figure 28.

as both hunter and biologist, in the functions of animal colors—and he regarded Thayer's obsession with concealment as both a nutty notion and an impediment to science. In fact, Roosevelt published, in 1911, a one-hundred-page monograph against Thayer's ideas: "Revealing and concealing coloration in birds and mammals," printed in a professional journal, the *Bulletin of the American Museum of Natural History.*

Not only did Roosevelt hold right on his side, and name value to his credit. America's former boss also carried a very big stick (and did not speak softly) as a polemical writer. Consider just one example from a private letter (though not much different in tone from many passages in his 1911 monograph), written to Thayer on March 19, 1912. (I confess that I also love this example as a testimony to the evolution of American politics and the nature of campaigning. In 1912, Roosevelt had split the Republican Party, formed his own Bull Moose group as a third-party insurgency against the incumbent W. H. Taft, thus effectively, if unintentionally, throwing the election to the Democratic candidate, Woodrow Wilson. Now, can you imagine any modern candidate, in the midst of such an effort, and just a month after the New Hampshire primary [please pardon my symbolic anachronism], taking time from the stump to write a long letter about natural history!)

> There is in Africa a blue rump baboon. It is also true that the Mediterranean Sea bounds one side of Africa. If you should make a series of experiments tending to show that if the blue rump baboon stood on its head by the Mediterranean you would mix up his rump with the Mediterranean, you might be illustrating something in optics, but you would not be illustrating anything that had any bearing whatsoever on the part played by the coloration of the animal in actual life. . . . My dear Mr. Thayer, if you would face facts, you might really help in elucidating some of the problems before me, but you can do nothing but mischief, and not very much of that, when conducting such experiments. . . . Your experiments are of no more real value than the experiment of putting a raven in a coal scuttle, and then claiming that he is concealed.

The second reason establishes the relevance of this example for a book on healing a misconceived gap between science and the humanities. Cheap shots come with the territory of human nature, but Thayer's opponents did not shrink from the philistine benefits of the old canard that only a person of artistic temperament, devoid of proper scientific training and understanding, could ever blunder so badly. For example, Teddy Roosevelt continued his attack in a statement that might have attracted even more attention in our litigious age: Thayer's errors, he opined, "are due to the enthusiasm of a certain type of artistic temperament, an enthusiasm also known to certain types of scientific and business temperaments, and which when it manifests itself in business is sure to bring the owner into trouble as if he were guilty of deliberate misconduct." Thomas Barbour, director of Harvard's Museum of Comparative Zoology (where I now work as professor and curator), stated:

"Mr. Thayer, in his enthusiasm, has ignored or glossed over with an artistic haze. . . . This method of persuasion, while it does appeal to the public, is—there is no other word—simply charlatanry however unwitting."

But this common charge just won't wash, and such personal branding by general caricature can only be called a cheap shot. Sure, Abbott Thayer became a classic victim of his own overexcitement and consequent extinction of good judgment. But I fail to see how this common human capacity correlates positively with art as a profession, or negatively with science as a calling. "True belief" can ensnare anyone in any activity—as Teddy Roosevelt at least had the decency to admit by including scientists among potential victims of such a temperament. Perhaps the rules of scientific procedure do act more effectively than the norms of some other lifestyles to discourage such unswerving commitment in the face of negative evidence. So one might anticipate a lower frequency of such behavior among professional scientists. (But I advance this hypothesis with only mild belief in its slight probability, and would certainly need to see gobs of hard data before reaching any conclusion.) In any case, the history of science remains chockablock with folks, including many people of great intellectual talent, who maintained, literally to their dying breath, pet theories and driving convictions just as uncompromisingly stated, and just as patently disproven (if only they had been willing to study the evidence), as Thayer's belief in the exclusivity of concealing coloration.

So why rehearse the old and sad tale of Thayer's chimerically invisible flamingos? Only to cast a few stones at scientists who went overboard in unfairly ascribing their empirically justified rebuttals to Thayer's artistic temperament? No, my method (at least in this case) betrays no madness; for I now ask you to back up and consider Thayer's first work in animal coloration, before the exclusivity of concealment captured his mind. In fact, Thayer not only made an important scientific discovery; he also reached his remarkable (and correct) conclusion by direct and conscious application of an artistic principle that had eluded all earlier scientists who had considered the same problem and failed because they had never encountered this conceptual key to resolution.

In a famous paper of 1896, titled "The Law Which Underlies Protective Coloration," Thayer solved the persistent problem of countershading. The colors of a countershaded animal are neatly graded to balance the effects of sunlight and shadow—usually dark on top grading evenly to a light (often truly white) belly. Biologists had long recognized the concealing value of countershading, but had assumed, before Thayer's work, that the effect arose by simple matching of colors. That is, a predator looking down upon the dark top of a potential prey would not distinguish the creature from the equally dark ground, whereas an enemy looking up would only perceive the white belly of potential prey, and then lose the animal as it blended into the bright sky.

But Thayer, as a trained artist who knew all the standard rules for depicting an illusion of three dimensions on a flat canvas, brilliantly recognized that countershading worked as nature's exploitation of a precise reversal—that is, by creating the illusion of an entirely two-dimensional object in a three-dimensional world. In short, Thayer recognized that countershading would conceal animals primarily by making them look flat, not mainly by matching their colors to their backgrounds.* Thayer knew this principle in his bones, and he then built decoy models of countershaded (and invisible) and inversely shaded (and doubly visible) birds to prove his point by striking demonstrations to skeptical biologists in the field (see figure 29).

Thayer convinced all doubters that the precise reversal between strength of coloration and intensity of illumination neatly cancels out all shadow and produces a uniform color from top to bottom. As a result, the animal becomes flat, perfectly two-dimensional, and cannot be seen by observers who have, all their lives, perceived the substantiality of objects by shadow and shading. Artists have struggled for centuries to produce the illusion of depth and roundness on a flat canvas; nature has simply done the opposite—she shades in reverse in order to produce the illusion of flatness in a three-dimensional world.

Contrasting his novel principle of countershading with older ideas about

---

*I learned to appreciate Thayer's point viscerally when I recognized the same principle behind one of the triumphs of modern architecture: the John Hancock Building, Boston's tallest. This glass tower rises high over Copley Square, right next to H. H. Richardson's magnificent Trinity Church. One would think that such a tall building, of such radically different style, would ruin and overwhelm the setting of one of Boston's finest, and basically late-Victorian to early-twentieth-century, public spaces. But one day I looked up and recognized that the Hancock Building, a very narrow parallelogram in plan view, has been cleverly sited so that, from nearly every crucial vantage point, one sees only the two dimensions of a single wall of glass (or just two of the sides as they meet at a highly obtuse angle, with no shadow cast across). And even though this wall rises more than sixty stories above the ground, the utter flatness renders the building effectively invisible, or at least entirely unobtrusive, if not actually enhancing as a blank "canvas" of sky to highlight the low buildings of Copley Square.

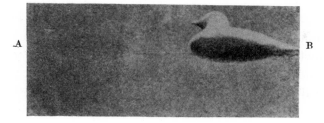

A   B

FIG. 1. Two bird-models, just alike except that the one on the left is counter-shaded, the other not, though covered uniformly with the very material of its background. This right-hand model, therefore, is *actually* as light below as above.

FIG. 2. Obliteratively-shaded bird-model, as in Fig. 1A, but inverted. (Somewhat side-lighted.)

Figure 29.

mimicry, Thayer wrote in his original statement of 1896: "Mimicry makes an animal appear to be some other thing, whereas the newly discovered law makes him cease to exist at all."

Thayer, in the joy of discovery, attributed success to his chosen profession and advanced a strong argument about the dangers of specialization and the particular value of "outsiders" to any field of study. He wrote in 1903: "Nature has evolved actual art on the bodies of animals, and only an artist can read it." And later, in his 1909 book, but now more defensively as his overextensions begin to attract valid criticism:

> The entire matter has been in the hands of the wrong custodi-
> ans. . . . It properly belongs to the realm of pictorial art, and can
> be interpreted only by painters. For it deals wholly in optical illu-

sion, and this is the very gist of a painter's life. He is born with a sense of it; and, from his cradle to his grave, his eyes, wherever they turn, are unceasingly at work on it—and his pictures live by it. What wonder, then, if it was for him alone to discover that the very art he practices is at full—beyond the most delicate precision of human powers—on almost all animals.

When I wrote my initial article on Thayer's discovery of countershading, I did not know that the artist's work on concealing coloration had also enjoyed an importance far greater than the merely abstract solution of an old problem in evolutionary biology. Thayer recognized the potential value of his findings in military camouflage, and he campaigned vigorously, both in America and in England (but with varying success), to convince our forces and allies to use his insights. He experienced much frustration, but eventually (if only posthumously, in World War II) gained the most precious reward of vital practical utility for good ideas, admittedly carried too far in his original intentions. I therefore end this section, and reinforce my basic claim for the benefits of *pluribus,* by quoting from two fascinating letters that I received, in response to my original article, from the former chief of naval camouflage, Lewis R. Melson, USNR. He wrote:

> Many years ago, I was summarily ordered to assume the responsibility for directing the efforts of the U.S. Navy's Ship Concealment and Camouflage Division, relieving the genius who had guided this effort throughout World War II, Commander Dayton Reginald Evans Brown. Dayton had perfected the camouflage patterns employed on all naval ships and aircraft throughout the war. . . . I found his theories and designs were based upon Abbott H. Thayer's earlier work in the field of concealment and camouflage. . . . Despite whatever everyone thought and thinks about Thayer's theories, both his "protective coloration" and "ruptive" designs were vital for concealing ships and aircraft.

Melson continued:

> All naval concealment and camouflage is designed for protection against the horizon in the case of shipping and either for con-

cealment against a sea or sky background, again at long ranges, for aircraft. Thayer's "Protective coloration" designs were outstanding for aircraft, light undersides and dark above. Ship concealment for temperate and tropical oceans employed the "protective coloration" designs, while "ruptive" or "disruptive" designs worked best against polar backgrounds.

Melson also taught me some history of camouflage during the two world wars. Despite our later and fruitful use in World War II, the U.S. Navy had originally rejected Thayer's proposal during World War I. However, Thayer had greater success in Britain, where his designs proved highly valuable during the First World War. Melson wrote:

> Thayer's suggestions . . . called for very light colored ships using broken patterns of white and pale blue. The intent of this pattern was to blend the ship against the background at night and in overcast weather. . . . These patterns proved very successful. HMS *Broke* was the first ship so painted and it was rammed twice by sister ships of the Royal Navy, whose captains protested that they had been unable to see *Broke*.

THE SCIENCE BEHIND POE'S GREATEST (AND ONLY) HIT

I begin with an old question in the general category of trivial pursuits with surprising resolutions: "What is the only work written by Edgar Allan Poe that appeared in a second edition during his lifetime?" Not "The Raven," which suffered the fate of its own refrain: "Nevermore." Not "The Fall of the House of Usher," which simply fell. Not "The Gold Bug," which sank as lead in the mighty waters (to quote Moses' assessment of Pharaoh's chariots). And not "The Murders in the Rue Morgue," unslated for resurrection until much later. The answer falls outside the experience of all but the most dedicated scholars of Poe's work: an apparently forgettable (and entirely forgotten) little textbook of 1839, titled *The Conchologist's First Book: or, A System of Testaceous Malacology, Arranged Expressly for the Use of Schools* (see figures 30 and 31 for Poe's own ID, if you doubt the claim and attribution). The first edition sold out in two months, leading to a second and enlarged version in 1840, and a

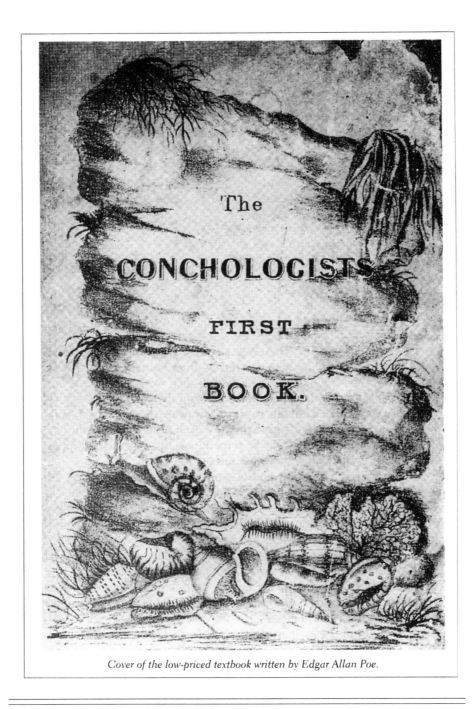

Cover of the low-priced textbook written by Edgar Allan Poe.

Figure 30.

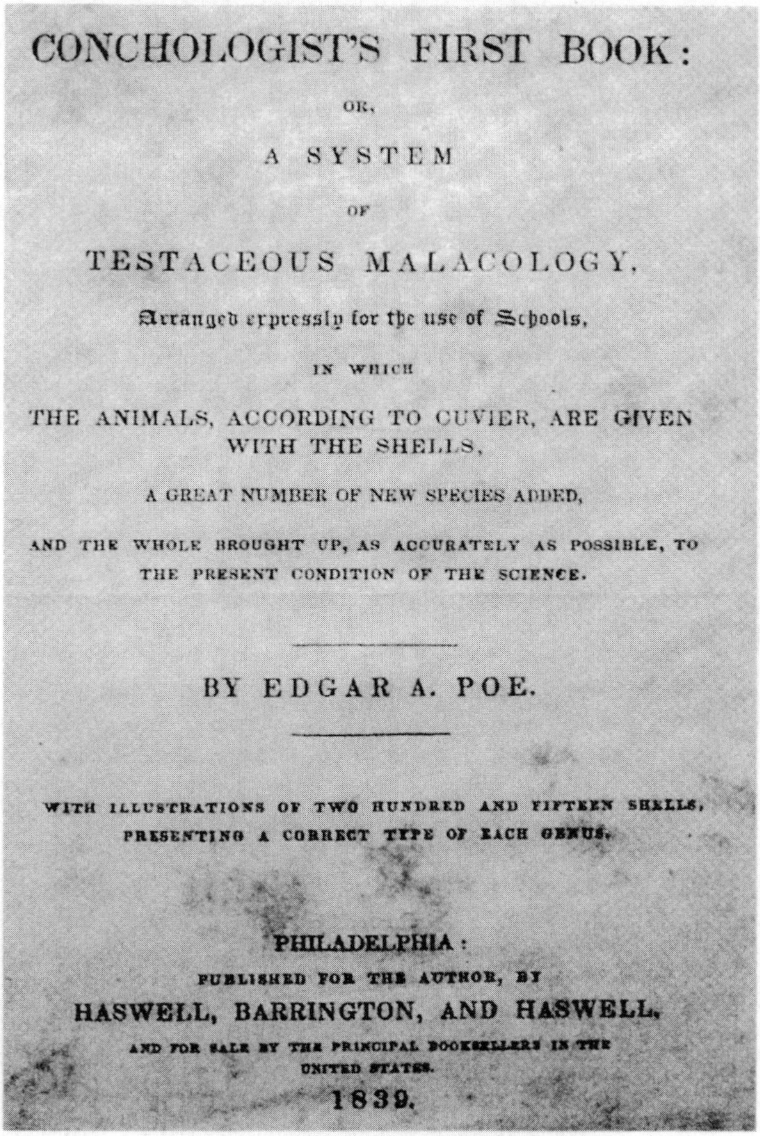

CONCHOLOGIST'S FIRST BOOK:

OR,

A SYSTEM

OF

TESTACEOUS MALACOLOGY,

Arranged expressly for the use of Schools,

IN WHICH

THE ANIMALS, ACCORDING TO CUVIER, ARE GIVEN WITH THE SHELLS,

A GREAT NUMBER OF NEW SPECIES ADDED,

AND THE WHOLE BROUGHT UP, AS ACCURATELY AS POSSIBLE, TO THE PRESENT CONDITION OF THE SCIENCE.

BY EDGAR A. POE.

WITH ILLUSTRATIONS OF TWO HUNDRED AND FIFTEEN SHELLS, PRESENTING A CORRECT TYPE OF EACH GENUS.

PHILADELPHIA:

PUBLISHED FOR THE AUTHOR, BY

HASWELL, BARRINGTON, AND HASWELL,

AND FOR SALE BY THE PRINCIPAL BOOKSELLERS IN THE UNITED STATES.

1839.

The title page lists Poe as the only author, although the book was a joint effort.

Figure 31.

third in 1845. Poor Poe probably only received a flat fee of fifty bucks for his role in the entire and (as we shall see) peculiar enterprise.

Everything about this curious work has presented a total puzzle and gnawing embarrassment to Poe scholars. First of all, no one has ever been able to figure out why he wrote the book, or ever got roped into such a project. Absolutely nothing in Poe's life and experience—he was, after all, the ultimate city boy and literary character—suggests any abiding interest, or even a sliver of concern, for natural history in any form.

The circumstances surrounding Poe's composition help to set a context, but, in another sense, have only deepened the mystery and enhanced the odor of disrespectability. Poe's friend Thomas Wyatt had, in 1838, published a lavish and expensive book on mollusk shells, retailing for eight dollars a copy. Sales were predictably slow, and Wyatt wished to produce a shorter and cheaper edition—especially since he made much of his living as a traveling lecturer on his generation's version of what would later be called the "Chautauqua circuit," serving local people eager for some education—the athenaeums, natural history clubs, and ladies' reading circles of America's isolated towns. Lecturers received fees for these mini-courses, but also supplemented their incomes by selling texts and pamphlets to accompany their lectures (just as modern musicians flog their CDs at intermissions between sets at coffee-house performances).

However, Wyatt's publishers understandably objected, citing a reasonable concern that their fancy edition would then become entirely unsellable. Wyatt, still wishing to proceed but fearing legal action should he publish the shorter version under his own name, sought a surrogate presence unlikely to attract litigation. At this point the plot thickens and the conventional shame accretes.

Literary scholars have been virtually unanimous in two characterizations of Poe's only reprinted work, with the second claim worse than the first. To begin with the lesser charge, nearly all critics brand *The Conchologist's First Book* as pure hackwork bearing no relation either to Poe's virtues or to his career. I surveyed all the standard biographies when I wrote my original article on Poe's greatest hit. The following sample presents a clear and uncontroverted consensus: F. T. Zumbach states that "it didn't bear even the slightest relevance to Poe's literary career." Julian Symons, an excellent writer of detective fiction as well as a literary biographer, writes that Poe "put his name to a piece of hackwork." David Sinclair describes *The Conchologist's First Book* as "a piece of shameful hackwork to which only desperation could have driven him." Jeffry Meyers labels the book as Poe's "grossest piece of hackwork."

The second, and more serious, charge of plagiarism seems more a matter of simple fact than of judgment. By current standards, Poe and his confrères would either be in jail or paying off a hefty fine. But, in 1840, copyright laws either lacked teeth or didn't exist at all—and Poe's actions, though indefensible, may not technically have been illegal.

The details bear accounting, for the import of my story hinges upon them. *The Conchologist's First Book* begins with a two-page "Preface," and I have no reason to doubt Poe's claim that he wrote this part all by himself. A four-page "Introduction" then follows—and now the trouble begins. Poe expropriated much of this text from the fourth edition (1836) of a British work by Captain Thomas Brown, the *Conchologist's Text Book*. Some biographers have claimed that Poe's entire "Introduction" is a paraphrase, if not a direct copy, of Brown. (F. T. Zumbach, for example, writes that Poe "copied from Brown almost word for word.") In fact, by my own comparison between the two books, only three of Poe's paragraphs (about one-fourth of the text) show extensive "borrowings." (Poe wins no exoneration thereby, for plagiarism, like pregnancy, does not increase in severity by degrees: beyond a point of definition, you either did or you didn't—and Poe surely did.)

The plot thickens with the next section of twelve plates. The first four, illustrating the parts of shells, are lifted *in toto* from Brown. No fuss, no pretenses, no excuses—just plain stolen. The subsequent eight plates, illustrating the genera of shells in taxonomic order, follow Brown in the more interesting pattern of back to front—that is, Brown's last plate becomes Poe's first (with considerable rearrangement, reorientation, and switching around of individual figures), and we then move up through Poe, and down in Brown, until Poe's last plate largely reproduces Brown's first.

Others have caught the pattern and even suggested that Poe and Wyatt were now consciously trying to hide their plagiarism. The actual reason is different and more interesting. (What could Poe and Wyatt be trying to hide anyway, after copying the first four plates exactly?) Brown's book follows the pedagogical scheme of the great French naturalist Lamarck, who always presented his discussions in the conventional order of a "chain of being," but from the top down, rather than the usual direction of bottom up. That is, Lamarck began with people and ended with amoebae, rather than the conventional vice versa. Brown followed Lamarck and therefore started with the most "advanced" mollusks, but Poe and Wyatt obeyed the usual convention and began with the most "primitive"—hence the reversed order of plates.

Charges of plagiarism surfaced in an 1847 article from the *Saturday Evening Post* in Philadelphia. Poe's response has often been quoted, but never taken seriously. I believe, however, that (despite some morbid self-pitying and exculpatory nonsense), Poe actually made a basically fair statement—and that the details of his defense can help us to solve all the puzzles of this old and troubling case. In particular, we can begin to understand why Poe, despite his utter ignorance of natural history, got the nod as Wyatt's reconfigurer; and, more important and surprising, why Poe (despite the indubitable plagiarism that no one should try to excuse) actually made a quite respectable and original contribution to the science, or at least to the teaching, of malacology (the study of clams, snails, and their allies)—the key point that requires the importation of a funny little fact from the history of science, an item that the literary critics never uncovered, thus explaining their failure to understand Poe's honorable role (and their consequent embarrassment at his evident culpabilities). Poe wrote to a friend about the charge of plagiarism:

> What you tell me about the accusation of plagiarism made by the "Phil. Sat. Ev. Post" surprises me. It is the first I heard of it. . . . Please let me know as many particulars as you can remember—for I must see into the charge—Who edits the paper? Who publishes it? Etc. etc.—about what time was the accusation made? I assure you that it is *totally* false. In 1840 [Poe is a year off here] I published a book with this title—The Conchologist's First Book. . . . This, I presume, is the work referred to. I wrote it, in conjunction with Professor Thomas Wyatt, and Professor McMurtrie of Ph[iladelphi]a—my name being put to the work, as best known and most likely to aid its circulation. I wrote the Preface and Introduction, and translated from Cuvier the accounts of animals etc. *All* school-books are necessarily made in a similar way. The very title page acknowledges that the animals are given "according to Cuvier." This charge is infamous and I shall prosecute for it, as soon as I settle my accounts with the Mirror.

Now note the four points that Poe here advances in explanation and excuse: first, that the work was composed by a committee, even though the

title page bore Poe's name alone; second, that he wrote the preface and introduction; third, that he also "translated from Cuvier the accounts of the animals"; and fourth, that "all school-books are necessarily made in a similar
way," presumably meaning that "borrowings" from previous work may be
regarded as *de rigueur* (as Poe then adds that the title page explicitly announces
a description of the creatures "according to Cuvier").

I will not defend the extent of "borrowing" in point four—surely, beyond
any permissible range, either then or now, and well into a realm that can only
be called plagiarism (Poe's consortium never mentions the name of their chief
source, poor Captain Brown). Neither can I entirely agree with the latter claim
of point two—for Poe expropriated at least a quarter of the "Introduction"
from Brown (although I believe he did write the "Preface" entirely by himself,
all two pages thereof).

When we read this preface, armed with basic knowledge about the history
of molluscan taxonomy, the more complex and favorable story begins to
emerge. This short statement emphasizes a single point: that *The Conchologist's
First Book* intends to do something different by describing *both* the shell and
the soft parts of each creature together. The claim seems awfully trivial, I
admit, and Poe does press his point only by the obtuse route of stressing an
expansion in terminology, from the traditional "conchology" (literally the
study of shells, as retained in the title) to "malacology" (or the study of the
entire organism—for the animals within the hard shells consist almost entirely
of soft parts, and the phylum's official name, Mollusca, derives from the Greek
word for "soft," as in our cognates *mollify* or *mollycoddle*). In any case, Poe
devotes his preface to this claim for expansion—and literary critics have never
granted the argument even a whiff of positive consideration. Poe writes:

> The common work upon this subject, however, will appear to
> every person of science very essentially defective, inasmuch as the
> relations of the animal and shell, with their dependence upon each
> other, is a radically important consideration in the examination of
> either. . . . There is no good reason why a book upon Conchology
> (using the common term) may not be malacological as far as it
> proceeds.

Poe then reinforces his intent by describing the new book's "ruling feature"—"that of giving an anatomical account of each animal, together with a

description of the shell which it inhabits." (Incidentally, a biography of Poe published in 1992 misses this point by failing to recognize the conceptual reform behind Poe's focus upon disciplinary names [malacology versus conchology]. The author writes that "Poe's boring, pedantic and hair-splitting Preface was absolutely guaranteed to torment and discourage even the most passionately interested schoolboy.")

But, in fact, although Poe's words now seem cryptic in the absence of a context that his contemporaries would have recognized, and that Poe fails to make explicit, his claims represent no mere airing of dry and inconsequential verbiage, but rather address the primary debate that had engaged generations of experts in the taxonomy and teaching of molluscan biology: Shall these creatures be ordered and classified by the shells alone, or should the soft anatomies within be considered as well (or even preferentially)?

Traditional classifications had used the shells alone (hence the description of the subject as "conchology"), but always apologetically. The great master, Linnaeus himself, had explicitly stated that a classification based on soft parts would be more "natural," but that he had utilized shells alone for the primary practical reason that collectors only retained these hard coverings, and that many genera, known only from empty shells collected at the shore, could not be defined from soft anatomy in any case. When Lamarck, in the next generation, presented the first major expansion and improvement in molluscan taxonomy since Linnaeus's efforts, he added many names and distinctions, but still based his system on the shells, not the soft parts.

To provide some sense of the salience and frustration of this issue among students of shells, consider this statement from the best popular book on the subject in English, *Elements of Conchology*, written by Emmanuel Mendes da Costa (a member of England's small community of Sephardic Jews) in the maximally auspicious year for contemplating change, 1776:

> This naturally leads me to the discussion of a subject of great
> debate among naturalists, which is, whether the methodical system
> or arrangement of testaceous animals should be formed from the
> animals themselves, or from their habitations or shells. The former
> method seems most scientifical; but the latter, from the shells, is
> universally followed, for many reasons: The vast number of species
> hitherto discovered, and the numerous collections made, exhibit
> only the shells or habitations, the animals themselves being scarcely

known or described. Of the shells we daily discover, few are fished up living; and the greater number are found on shores, dead and empty. . . . How is it possible then to arrange a numerous set of animals by characters or parts [that is, by the soft anatomy], we can with difficulty, if ever, get acquainted with, in the far greater number of the species we collect or discover?

Thus, as Wyatt, Poe, and friends planned *The Conchologist's First Book,* they decided to introduce a substantial reform by describing the soft parts of each animal along with the shell. However, they had to face the serious problem that all works in English, both technical and popular, treated mollusks by the traditional method of discussing the shells alone. Wyatt's original and expensive version described only the shells, as did Brown's volume, the source of Poe's plagiarism. As an example, a popular work on mollusks, published in 1834 by Mary Roberts, begins by separating the study of animal and shell, and by defending a treatment based upon shells and ignoring the animals entirely: "The elegant science of Conchology, my friend, comprises the knowledge, arrangement, and description of testaceous animals; a science, according to Linnaeus which has for its basis the internal form and character of the shell, and is totally independent of the animal enclosed within the calcareous covering." And Thomas Brown added, in 1836, "It is upon the exclusive shape of the shell, not the animal inhabitant, that the Linnaean arrangement of Conchology is formed."

But Wyatt persisted in his desire for innovation, heeding the following lament written by Mendes da Costa in 1776:

I am well aware of the arguments alleged against it [that is, the classification by shells that Da Costa actually used in his book], *viz.* that, as long as we study only the very shells, those empty habitations, those spoils or remains only of the animals, the present sole objects of our researches and collections, we consider these beings but partially, or with a side view. There is more to be required. The animals that inhabit them should certainly guide us in our methodical arrangements.

Where, then, could Wyatt find the data on soft anatomy to integrate with the familiar and conventional descriptions of shells? Now, and finally, we can understand and appreciate Poe's vital, indeed necessary, role in an admittedly dubious, and undoubtedly plagiarized, enterprise that, nonetheless, achieved something worthy, original, and heretofore unrecognized by Poe's literary admirers and critics. As stated above, no English publication could provide the required information. In the early to mid-nineteenth century, French science, centered at the Muséum d'Histoire Naturelle in Paris, the professional home of both Lamarck and Cuvier, led the world in taxonomy and natural history. Adequate information about the soft anatomy of mollusks could only be found in the primary and technical literature, written in French!

Now Poe may not have known a mollusk from a martini, but he was certainly fluent in French—probably the only member of Wyatt's circle with sufficient expertise in this essential ingredient of any effort to link the shells and soft parts of mollusks in a popular English presentation. Poe's actress mother had died when Poe was only two, and he had been raised in the home of an intermittently wealthy Richmond businessman, John Allan (from whom Poe took his middle name, although he was never formally adopted). Poe lived in England and Scotland for five crucial years (1815–1820), where he received a classical education in rigorous schools, including a thorough grounding in French.

In other words, and in conclusion, I think that Poe did exactly what he said—and that no one else in Wyatt's crowd could have consummated this important project. Poe translated the descriptions of molluscan soft parts from Cuvier's French and then united this information with the traditional accounts of shells. Thus, *The Conchologist's First Book* presented an important, and widely desired, educational reform by linking, for the first time in a popular English book, the shells of mollusks with the bodies housed within and responsible for the elegant constructions—an innovation well meriting a reprint or two! And Edgar Allan Poe played a crucial role, absolutely essential (given Wyatt's limited contacts and resources) for the successful completion of this reform. Thus, Poe served science well because he possessed the humanist's skill of fluency in French. *E pluribus,* a better *unum.* Borrow one of the fox's skills, and advance the cause of the hedgehog.

9

The False Path of
Reductionism and the
Consilience of Equal Regard

A CLASSICAL PROGRAM OF
HUMANE REDUCTIONISM: A BEST
TRY AT A LOGICAL IMPOSSIBILITY

HUMAN SAGAS AND PRIMAL TALES OFTEN DEPICT OUR DEEPEST EMOTIONS and most practical needs as polar opposites that either dwell in tension within us or vie for domination as personified beings of the outside world (super-heroes and villains of modern comic books as pop versions of ancient gods and devils, for example): kill for personal gain or sacrifice for national salvation; dance till death (shop till you drop), or study to blindness. As a scientist and natural historian, I especially feel the strong personal pull of opposition between an irreducible fascination (amounting to love) for every little detail of natural variety, and a great yearning (amounting to thrill) at the prospect that one common mode of explanation, one governing principle, might just make sense of all the glorious diversity. Why else would a natural historian feel compelled to write books about these irreconcilable feelings, anchoring the efforts in the obvious academic "excuse" (and quite reasonable context) for a legitimate historical and philosophical disquisition: the relationship between science and the humanities. If the yearning were not so strong and pervasive, and the range of honorable solutions not so broad, I doubt that my close colleague E. O. Wilson could write a book on the same subject (*Consilience*, Knopf, 1998), yet reach such an opposite conclusion within our common conviction that (to quote Wilson again, as on page 3) "the greatest enterprise of the mind has always been and always will be the attempted linkage of the sciences and humanities."

Clearly, no sensible person will advocate a pure extreme—that is, either the conceptual version of "one size fits all," the discovery of the single mantra, the *Om* of God's name, the abracadabra of existence (releasing the djinn of infinite wishes); or the anarchic alternative that each natural item inhabits its own ineffable space of gorgeously unique solitude, without even a gossamer thread of connection to any other, no sense of any order or coordination at all, not a conceptual higher or lower, or even a geometric nearer or farther away. We all want to enjoy the differences, yet find some meaningful order in the totality. In this primal sense, everyone appreciates both the fox's flexible range and the hedgehog's steady effectiveness.

But the social traditions and conventional intellectual formulations of modern Western science have favored an emphasis upon searches for unification through reduction to a limited number of highly generalized and interconnecting principles regulating fewer forces and smaller constituent particles, even leading to a bit of intended self-parody in talk about searches for a "final theory" or for the truly foundational acronym of TOE, or "theory of everything." I have already quoted the classical scientific assertion of this faith— Galileo's famous statement that the "grand book" of the universe "is written in the language of mathematics, and its characters are triangles, circles, and other geometrical figures." But perhaps the most scientifically interested and savvy of Victorian poets, Alfred, Lord Tennyson, made the same point even more powerfully, if from his own magisterium of metaphor:

> One God, one law, one element.
> And one far-off divine event,
> To which the whole creation moves.

(from the epilogue of his most famous poem, *In Memoriam*).

Ed Wilson, who has presented, in *Consilience,* the most eloquent recent defense of this synthesizing preference in his plea for respectful unification of the sciences and humanities, borrows a lovely phrase from the physicist and historian of science Gerald Horton to describe this emphasis upon a grand unification stretching through the "hardest" science of tiny constituents to the crown of the humanities, or from physics through biology to the social sciences, arts, and human ethics:

> The Ionian Enchantment . . . It means a belief in the unity of the sciences—a conviction, far deeper than a mere working proposition, that the world is orderly and can be explained by a small number of natural laws. Its roots go back to Thales of Miletus, in Ionia, in the sixth century B.C. . . . The Enchantment, growing steadily more sophisticated, has dominated scientific thought. In modern physics its focus has been the unification of all the forces of nature—electroweak, strong, and gravitation—the hoped-for consolidation of theory so tight as to turn the science into a "perfect" system of thought, which by sheer weight of evidence and logic is made resistant to revision. But the spell of the Enchantment extends to other fields of science as well, and in the minds of a few it reaches beyond into the social sciences, and still further, as I will explain later, to touch the humanities.

Wilson does not shy away from granting this traditional dream of unification both its usual direction of subsumption, and its conventional name of "reductionism": the program of practical research (generally buttressed by a belief, or at least a suspicion, about the actual construction of material reality) that attempts to break the most complex phenomenology (of living, cognitive, and social systems) into constituent units, all ultimately subject to explanation by the unifying physical laws regulating these basic components. Wilson also (pages 58–60) ends his statement by affirming science's admirable practice of skepticism, with special severity directed at one's fondest hopes:

> The cutting edge of science is reductionism, the breaking apart of nature into its natural constituents. The very word, it is true, has a sterile and invasive ring, like scalpel or catheter. Critics of science sometimes portray reductionism as an obsessional disorder, declin-

ing toward a terminal stage one writer recently dubbed as "reductive megalomania." That characterization is an actionable misdiagnosis. Practicing scientists, whose business is to make verifiable discoveries, view reductionism in an entirely different way: It is the search strategy employed to find good points of entry into otherwise impenetrably complex systems. Complexity is what interests scientists in the end, not simplicity. Reductionism is the way to understand it. . . .

Behind the mere smashing of aggregates into smaller pieces lies a deeper agenda that also takes the name of reductionism: to fold the laws and principles of each level of organization into those at more general, hence more fundamental levels. Its strong form is total consilience, which holds that nature is organized by simple universal laws of physics to which all other laws and principles can eventually be reduced. This transcendental world view is the light and way for many scientific materialists (I admit to being among them), but it could be wrong. At the least, it is surely an oversimplification. At each level of organization, especially at the living cell level and above, phenomena exist that require new laws and principles, which still cannot be predicted from those at more general levels.

Wilson revives an old word, *consilience,* invented by the great English philosopher of science William Whewell in 1840 (I shall provide definitions and analyze Wilson's misunderstandings of Whewell's intentions on pages 200–215). I regard *consilience* as a lovely and deserving term that never caught on in the "natural selection" of English vocabulary. The word literally designates the validation of a theory by the "jumping together" of otherwise disparate facts into a unitary explanation. Wilson revives Whewell's word to describe the most powerful putative result of reductionism's triumph: the simplification and gathering together of vast ranges of phenomena by their successive subsumption under laws governing constituent parts, right down to the physics of basic constituents. In Wilson's dream—the full scale of his Ionian Enchantment—this range of consilience, with reductionism as the explanatory guide, will extend from the physics of elementary particles, to biological and social systems, right up and through the greatest traditional divide (science and humanities) into the arts and ethics as well (pages 221–222)—an extension and alteration of "consilience" quite inconsistent with Whewell's intentions and basic beliefs about relationships between science and other parts of human life.

> The central idea of the consilience world view is that all tangible phenomena, from the birth of stars to the workings of social institutions, are based on material processes that are ultimately reducible however long and tortuous the sequences, to the laws of physics. . . . The strategy [of reductionism] that works best in these enterprises is the construction of coherent cause-and-effect explanations across levels of organization. Thus the cell biologist looks inward and downward to ensembles of molecules, and the cognitive psychologist to patterns of aggregate nerve cell activity. . . . No compelling reason has ever been offered why the same strategy should not work to unite the natural sciences with the social sciences and humanities. The difference between the two domains is in the magnitude of the problem, not the principles needed for its solution.

In fairness, for I do regard the logical rebuttal of Wilson's vision as a strong argument for embracing my alternative form of journey toward a conjunction of the sciences and humanities, I do appreciate his chosen central metaphor for consilience and reductionism—for Wilson's image reverses the usual geometric picture, thereby rebutting one of the worst collateral implications of classical reductionism. In the usual view, we rank the sciences supposedly subject to reduction in a hierarchy of worth, with adamantine physics on top, and squishy subjects like sociology and psychology on clay feet below. (I must also confess a personal aversion to this picture because, as a paleontologist, I may work with literally hard objects, but my profession surely resides near the squishy conceptual terminus of this continuum!) Thus, the littlest with the mostest generality (and mathematics) is best (particle physics), and the biggest, with a maximally confusing amalgam of different things explained by fewest organizing principles (complex ecological systems, for example), is worst.

At least Wilson reverses this imagery with a comparison to the tale of Theseus and the Minotaur, with the center of the labyrinth not as the starting point for all buildups to larger things nearer the periphery, but rather (and in reverse) as the most difficult goal of final accumulation, with a consilient gain added at each turning from the outside in. That is, you begin with basic particle physics as you enter the labyrinth, and then follow a pathway of ever greater complexity until you reach the most difficult problem at the center (also requiring the not inconsiderable task of slaying a nasty anthropophagous creature with a bull's head once you arrive). Of course, you can never get out (whatever success you enjoy at the center) unless you carefully lay Ariadne's thread along your path (corresponding to the intellectual process of continuous subsumption in reductionist explanation), and then trace your conquest at the complex center back through all the levels of ever more general analysis, based on ever more basic constituents, until you reach the physics of the periphery.

This striking image, while rejecting the conventional order of worth (after all, Wilson and I both work as evolutionary biologists, and should be equally loath to regard particle physics as a supreme source), also serves Wilson well in permitting him to make the important point, which I loudly applaud, that the reductive process of consilience can proceed equally well and effectively in either direction, and need not start with quarks, leaving *Quercus* (oak trees) unaddressed until all the levels below fall into place (pages 73–74):

> Theseus is humanity, the Minotaur our own dangerous irrationality. Near the entrance of the labyrinth of empirical knowledge is physics, comprising one gallery, then a few branching galleries that all searchers undertaking the journey must follow. In the deep interior is a nebula of pathways through the social sciences, humanities, art, and religion. If the thread of connecting causal explanations has been well laid, it is nonetheless possible to follow any pathway quickly in reverse, back through the behavioral sciences to biology, chemistry, and finally physics. . . . To dissect a phenomenon into its elements . . . is consilience by reduction. To reconstitute it, and especially to predict with knowledge gained by reduction how nature assembled it in the first place, is consilience by synthesis. That is the two-step procedure by which natural scientists generally work: top down across two or three levels of organization at a time by analysis, then bottom up across the same levels by synthesis. . . . There is another defining character of consilience: It is far easier to go backward through the branching corridors than

to go forward. After segments of explanation have been laid one at a time, one level of organization to the next, to many end points (say, geological formations or species of butterflies) we can choose any thread and reasonably expect to follow it through the branching points of causation all the way back to the laws of physics. But the opposite journey, from physics to end points, is extremely problematic. As the distance away from physics increases, the options allowed by the antecedent disciplines increase exponentially. Each branching point of causal explanation multiplies the forward-bound threads. Biology is almost unimaginably more complex than physics, and the arts equivalently more complex than biology.

I have no desire to engage in the crudest form of pop psychology, or psychobabble, but I have often wondered why the dream of unification (in our horrendously messy, yet so wondrously multifarious world) holds such power over the scholarly mind. I should, of course, begin with the honest (and rather obvious) admission that such senses of puzzlement tend to increase when the puzzled person doesn't share the belief that seems so patent and potent to so many respected others. I find nothing either viscerally or intellectually appealing in such neatly and symmetrically honed structures with no rough edges or outlying separate islands (even substantial continents) that don't connect by intimate physical ties (but may well conjoin in more interesting nonphysical or nonlogical senses better expressed by metaphor or by truly different ways of honoring some higher and eminently worthy commonality). After all, I have written this book, and used the fox and hedgehog rather than Ariadne's labyrinth as its defining image, because I want the sciences and humanities to become the greatest of pals, to recognize a deep kinship and necessary connection in pursuit of human decency and achievement, but to keep their ineluctably different aims and logics separate as they ply their joint projects and learn from each other. Let them be the two musketeers—both for one and one for both—but not the graded stages of a single and grand consilient unity.

Nietzsche's famous distinction of the Apollonian (critical-rational) from the Dionysian (creative-passionate) aspects of human motivation may help us to understand two extreme bases for powerful attraction to the ideal of unity. Wilson himself claims illumination from the most Apollonian source of this Enchantment, as expressed in the great secular and intellectual movement of the eighteenth century, the Enlightenment—an episode that he continues to regard as an apogee, albeit an ultimate failure (and for a definite reason), of Western confidence in the power of reason to improve our lives by ensuring continual progress, both factually and morally. Wilson writes (page 15):

> The dream of intellectual unity first came to full flower in the original Enlightenment. . . . A vision of secular knowledge in the service of human rights and human progress, it was the West's greatest contribution to civilization. It launched the modern era for the whole world; we are all its legatees. Then it failed.

Wilson ascribes the failures of the Enlightenment to two major reasons, one internal and the other external. For the crumbling from within, these great thinkers could not carry out their rationalist program because the science of their day could not reach sufficiently "upward" in its consilience to explain the complex parts of nature most central to the goal of making our social and economic lives more rational and humane—starting from the human brain and moving upward to social organization and history. Of the Marquis de Condorcet (one of my favorite historical characters as well), who maintained his vision of human perfection even as zealots in the most radical phase of the French Revolution (that he had supported so ardently at its hopeful beginnings) hunted him to his death, Wilson writes (page 21): "His serene assurance arose from the conviction that culture is governed by laws as exact as those of physics. We need only understand them, he wrote, to keep humanity on its predestined course to a more perfect social order ruled by science and secular philosophy."

Wilson also attributes the failure of later movements for intellectual unification, when based on the same rationalist spirit, to a similar inability to explain complex levels (beginning with the human brain) in scientific terms, thereby precluding the incorporation of these essential subjects into the reductionistic consilience. Of logical positivism, the twentieth-century philosophical movement best developed by a group known as the Vienna Circle (at least until Hitler's policies forced the dispersion—or encompassed the death—of several key Jewish members), Wilson writes (page 69):

> Logical positivism was the most valiant concerted effort ever mounted by modern philosophers. Its failure, or put more generously, its shortcoming, was caused by ignorance of how the brain

works. That is my opinion of the whole story. No one, philosopher or scientist, could explain the physical acts of observation and reasoning in other than highly subjective terms.

For the defeat of the Enlightenment from without, Wilson notes the powerful forces of human mentality and tradition that find secular reason just a bit "bloodless" and that seem to need (and desperately seek) the thrill of visceral oneness, imposed from above by an authoritarian mystery, worthy of worship. Contrasting the Enlightenment's deistic "take" on religion with more conventional Western views, for example, Wilson writes (page 36):

> The fatal flaw in deism is thus not rational at all, but emotional. Pure reason is unappealing because it is bloodless. Ceremonies stripped of sacred mystery lose their emotional force, because celebrants need to defer to a higher power in order to consummate their instinct for tribal loyalty. In times of danger and tragedy especially, unreasoning ceremony is everything. There is no substitute for surrender to an infallible and benevolent being, the commitment called salvation. And no substitute for formal recognition of an immortal life force, the leap of faith called transcendence.

And yet Wilson also (and frequently) invokes the opposite, strongly Dionysian and basically Romantic, rationale for unification as well (using Romantic in a historical and technical sense to define the movement, exalting the primacy of powerful feeling and the innate nature of our emotional needs, that, in the late eighteenth and early nineteenth centuries, became so popular among Western literati and intellectuals following their disenchantment with earlier Enlightenment ideals). Wilson usually cites reduction through his consilient chain (back to principles regulating the more successfully established sciences of smaller components) to support particular claims for explaining more-complex systems, from the brain to human society and on to the arts, ethics, and religion.

Nonetheless—and at several crucial points—he seems to slip into the following basically Romantic argument: How can we validate an ethical principle generally not regarded as enmeshed within a consilient chain that could receive factual affirmation by reduction to a more mature science nearer the perimeter of Ariadne's labyrinth? Wilson seems to locate an alternative form of "objective" validation—and I don't know what else to call such a claim but "romantic"—in a principle of consonance with evolved preferences of the human mind, as expressed in the resonance of an ethical precept with a strong and innate feeling within our common nature. We may, by cultural tradition based on misunderstanding of our motives and mental processes (or even just to win greater success in compelling obedience), try to validate the ethical principle in religious or other nonlogical and nonscientific terms, but the "true" basis of conformity with our evolved nature and being remains preeminent, if unaddressed (and even if entirely unappreciated):

> The empiricist view in contrast, searching for an origin of ethical reasoning that can be objectively studied, reverses the chain of causation. The individual is seen as predisposed biologically to make certain choices. By cultural evolution some of the choices are hardened into precepts, then laws, and if the predisposition or coercion is strong enough, a belief in the command of God or the natural order of the universe. The general empiricist principle takes this form: *Strong innate feeling and historical experience cause certain actions to be preferred; we have experienced them, and weighed their consequences, and agree to conform with codes that express them. Let us take an oath upon the codes, invest our personal honor in them, and suffer punishment for their violation.* The empiricist view concedes that moral codes are devised to conform to some drives of human nature and to suppress others.

But the last line exposes the evident dilemma: if human experience, based on consonance with our evolved nature, has led us to prefer certain behaviors (subsequently defined as morally correct and embodied in codes of action), then how can ethical systems be validated within the reductionist chain—that is, as part of factually "true" knowledge, albeit of the most complex and difficult systems—if, at the same time, we must admit that these preferences "conform to some drives of human nature" but "suppress others"? Aren't the preferences that we suppress just as "natural," and just as factually evolved, as the ones we have favored? How, then, can we choose, unless we step outside the reductionist chain of scientific fact and admit a different kind of basis for moral validation? Presumably, Wilson would argue that the preferred drives can be factually validated at a still higher level of rules for human cultural organization, rather than biological predispositions of the evolved brain.

But, at this crucial point in his entire consilient system, Wilson can offer little beyond a statement of hope, or even faith, that of the choices sanctioned by human mental possibilities, we will select the path supported by our "best" instincts for democracy and toleration—a path that will also provide the maximal probability for prolonged survival (in decency) as a species respectful of its planetary home. And if an essential component of the Romantic perspective lies in granting such a defining power to our deepest emotional realities, equating "truth" with their very existence, then I can only regard Wilson's ultimate basis for preferences among "natural" drives as an even greater departure from the chain of consilience, and a stronger move toward a basically romantic form of validation: that is, within the natural set, choose the alternatives that exalt our "nobler" propensities, even if a full gamut of other possibilities remains factually accessible within us. But I do not think that "nobility," in this sense, can possibly claim a factual or scientific definition in any consilient chain:

> How can the moral instincts be ranked? Which are best subdued and to what degree, which validated by law and symbol? How can precepts be left open to appeal under extraordinary circumstances? In the new understanding can be located the most effective means for reaching consensus. No one can guess the form the agreements will take. The process, however, can be predicted with assurance. It will be democratic, weakening the clash of rival religions and ideologies. History is moving decisively in that direction, and people are by nature too bright and too contentious to abide anything else. And the pace can be confidently predicted: Change will come slowly, across generations, because old beliefs die hard even when demonstrably false.

THE TWO CHIEF FALLACIES OF REDUCTIONISM AND THE ORIGINAL MEANING AND INTENT OF CONSILIENCE

REASSERTING A DIFFERENT APPROACH TO THE COMMON GOAL OF MAXIMAL UNITY BETWEEN THE SCIENCES AND HUMANITIES

One might propose two kinds of solutions in principle—each in pursuit of the same worthy goal but one (perhaps) right and the other unworkable—to a common dilemma that may well bear the name of a famous exemplar: the Ugly Duckling. How shall an apparent (and ungainly) misfit win acceptance among its fellows, the common goal of both approaches? We might, if the Ugly Duckling represents the humanities and his fellows the sciences, try to convince the ordinary host that their clumsy brother really belongs to the same stock, and that they have held him in disregard only because they have not classified the differences among their members correctly. They saw the Ugly Duckling as big and awkward because they didn't realize that their unity forms a consilient chain, with sleek and well-coordinated little ducklings at one end and awkwardly complex cygnets at the other. Each end exhibits great virtues, and neither should be judged better. The sleek end may revel in its status as an ultimate source of explanation for the increasingly complex organization of its constituents up the chain of the full series; but the Ugly Duckling, at the big and clumsy end, may take just as much pride as the most complex configuration, and most difficult to appreciate or understand. But why engage in such silly arguments at all if, as Mr. Pope of my initial story in chapter 1 stated, albeit in poetic metaphor, "all are but parts of one stupendous whole; whose body nature is, and God the soul."

This first solution embodies Wilson's proposal in his volume *Consilience*. But we also might, in a second solution promulgated in the present book, try to convince the ordinary host that their ungainly ally really belongs to a different "natural kind," and that his apparent clumsiness only indicates their failure to understand his disparate but equal excellence. In fact, once they invoke their goodwill to sort out the legitimate differences, they will realize the enormous weight of common interests, and recognize two vital powers and pleasures that their new insight about inherent difference can only enhance. First, they can have ever so much fun together, horsing around in the pond and swapping stories, because, after all, swans do share many features with ducks, so the commonalities ensure mutual comprehension while the differences enrich the stories. Second, as the saying goes, "we two form a multitude"—and what a power of influence and respect can emerge from a union rooted in common goals equally backed by different skills contributed toward their realization.

I do apologize to you, Mr. Andersen, for this ludicrous expropriation, but I also thank you for a direct inspiration, literally rooted in *genius loci.* For my wavering intention to write this book congealed when I received an invitation to your natal town of Odense to present, as a scientist, the "first annual"— don't you love the optimism in such a description!—Hans Christian Andersen lecture at your city's University of Southern Denmark. I chose for my title "The Necessary Role of Story-Telling in the Sciences of Natural History." *Vive la différence,* and the potential for fruitful union. (The second part of this previous sentence, come to think of it, is the point of the first, *n'est-ce pas?*)

I can summarize my reasons for rejecting Wilson's solution, while reasserting the case for my alternative, by critiquing both the meaning and applicability of the two key words and concepts that define and embody his analysis: *reductionism* and *consilience.* The closing three sections of this part of chapter 9 will present the full case, now only epitomized below:

1. I believe that reductionism—a powerful method that should be used whenever appropriate, and that has been employed triumphantly throughout the history of modern science—must fail as a generality (both logically *and* empirically) for two crucial and entirely different reasons, each relevant to a central, but different, aspect of Wilson's case:

(i) Within the legitimate magisterium of science, I do not believe that reductionism can come even close to full success as a style of explanation for levels of complexity (including several aspects of evolutionary biology, and then proceeding "upward" in intricacy toward cognitive and social systems of even greater integration and interaction) for two basic reasons that allow these subjects to remain fully within the domain of factual and knowable science, but that require additional styles of explanation for their resolution. I will explicate (on pages 221–232) what can only sound like mumbo-jumbo in this brief statement, but I must at least record these two basic reasons here. First, *emergence,* or the entry of novel explanatory rules in complex systems, laws arising from "nonlinear" or "nonadditive" interactions among constituent parts that therefore, in principle, cannot be discovered from the properties of parts considered separately (their status in the "basic" sciences that provide the fundamental principles of explanation in classically reductionist models). Second, *contingency,* or the growing importance of unique historical "accidents" that cannot, in principle, be predicted, but that remain fully accessible to factual explanation after their occurrence. The role of contingency as a component of explanation increases in the same sciences of complexity that also become more and more inaccessible to reductionism for the first reason of emergent principles. A quark may not owe its defining characteristics to accidents of history, but only to lawful rules of natural order; however, the emergence of *Homo sapiens* as a small population in a certain place at a certain time (Africa within the past 200,000 years), while disobeying no natural law, and while helpfully elucidated by several properties of these laws, cannot be explained meaningfully without emphasizing the formative role of historical contingencies that, in principle, do not flow predictably from laws of nature (even though such contingent events cannot confute these laws either).

(ii) Beyond the legitimate magisterium of science, and in key fields of the humanities that must be brought into the consilient chain of explanation for Wilson's model to prevail, the basic inquiries, desiderata, and modes of resolution preclude, logically and in principle, any full or even vaguely satisfactory explanation by the factual methods of empirical science at any level in any reductionistic chain. I do not, of course, deny that factual questions of importance apply to all these fields in the humanities. Surely we may usefully pursue the anthropology of morals as a question of relative frequency among independent cultures, or the psychology of aesthetics in optical terms, for example. We can surely determine that a great majority of human societies have preferred one moral code over another, and we may even be able to devise a satisfactory evolutionary explanation for the decision. But the magisterium of ethics asks a very different primary question, unaddressed (and unaddressable) by such interesting and important factual data: What moral code *ought* we follow? What ethical duties define a life well lived? How, in a purely logical sense, can the factual anthropology of morals resolve, or even usefully help to adjudicate, such a question? In a hypothetical example previously cited (page 142), if we discover that a majority of human cultures have favored infanticide under certain conditions, and that such a practice arose for good

Darwinian reasons, shall we then claim that we have resolved the question of the rightness of such a practice with a "yea"? (I would say, to the contrary, that we have, at most—and such knowledge should be treasured as highly useful—only learned that our job will be more difficult as we try, for moral reasons explicated and validated by modes of reasoning outside factual science, to eliminate this ancient and widespread practice.)

2. I am delighted that Wilson has rescued, and restored to prominence, one of my favorite obscure words, indeed my longtime leading candidate for a term that should have stuck, but suffered apparent extinction instead—at least until Wilson lit the Phoenix's pyre. But, although Wilson correctly interprets Whewell's original application of consilience in the specific domain of scientific knowledge, he also, and in a strikingly ironic manner, then extends "consilience" into a name for a program that directly contradicts the larger worldview of England's greatest mid-nineteenth-century historian and philosopher of science. Whewell's own view of distinct magisteria of knowledge coincides with the views supported in this contrary brief, and not with the single reductionistic chain championed by Wilson when he borrowed Whewell's word for his central premise and book title.

Of course, terms are as labile and as subject to evolution as organisms, so Wilson may surely propose such an extension from Whewell's original application to scientific knowledge into a far broader claim for further consolidation by encompassing the traditional subjects and questions of the humanities within the same explanatory structure as well. Still, substantial irony inheres in the peculiar circumstance that an explicit restriction placed by Whewell upon his own term *consilience* happens to embody the central argument for the primary failure of Wilson's program.*

---

*As I confessed a personal reason for citing Andersen above, I would not be following norms of proper disclosure (despite my personal distaste for "confessional" writing) if I did not include a few lines about my collegial history with E. O. Wilson. First of all, and to admit something in myself that I can only deem petty, I was a bit peeved when Wilson chose *Consilience* as the title to his 1998 book. In my own work on the history and philosophy of science, with my particular focus on Darwin and the Victorian period in general, I had studied Whewell's concept of consilience, and had used the term and idea prominently in two papers, as the centerpiece for describing Darwin's historical methodology (Gould, 1986) and in defending my own style of empirical documentation in a major monograph on the taxonomy of a particularly difficult group of land snails (Gould and Woodruff, 1986). I

continued

thought that I was the only living evolutionary biologist who had ever discovered and used Whewell's term. (I guess I mention this to give readers ammunition for suspecting a personal motivation that might legitimately be called mean-spirited if you feel that I have gloated a bit too much in arguing that Wilson misconstrues Whewell's motives and intentions.)

It is also scarcely a secret that Ed Wilson and I have had our disagreements on some theoretical issues in evolutionary biology, centered upon the roles of adaptation and the applications of traditional Darwinian arguments to certain forms of human social behavior. Since people assume that intellectual heat must breed emotional fire, I believe that many people assume hostility between us. I can only say that I have never experienced any personal difficulties or animosity with Ed Wilson, that I do not remember a harsh word ever passing between us in any verbal exchange, and that our relationship has always been entirely collegial and respectful, although we have (I suspect for reasons of common temperament as notorious loners) never became friends in a personal sense.

I also wish to comment on one incident that continues to fill me with chagrin, although I believe that both Ed and I acted entirely honorably—with *far more* honor to him, as I have long regretted an action I did *not* take. At a meeting of the American Association for the Advancement of Science, Ed and I participated in a tough and wide-ranging session of criticism on his theory of sociobiology. In these times of more radical student politics, a group of juvenilistic ideologues (I will not dignify their actions with the name of any serious political or scientific theory), arguing that sociobiology spread racism (a nonsensical charge since the theory deals with putative human universals, and not with the causes of geographically based variation, the pseudoscientific substrate for racism), rushed the stage, and "demonstrated" with chants and charges. One student, yelling "Racist Wilson, you're all wet," took a cup of water and poured it over Wilson's head. The group then left the stage and the hall. (I was seated right next to Ed and got pretty wet myself.)

The incident, ugly enough already, for obvious reasons both general and specific, became even uglier because Wilson, at the time, had his ankle in a cast and would not have been able to defend himself physically, had the necessity arisen. I praise Wilson over me for two reasons. I took the microphone and denounced the protestors who had so sullied and destroyed our attempt to present a serious and respectful, if intellectually tough, critique of sociobiology. I cited one of their own supposedly canonical documents against them— Lenin's pamphlet describing "left-wing communism" (a similar movement of his time, based on silly show rather than serious theory) as "an infantile disorder." Ed simply wiped himself off and continued his talk. His silent dignity beat my impassioned outrage by an order of magnitude. He was also the target of their attack and had real reason for fear.

My chagrin requires a personal confession as well: I have never, in my adult life, hit another human being (well, maybe I once gave one of my kids a really light *potch* on the *tochas* after a particularly outrageous outburst). But I would give a great deal to have that one moment back—to alter contingent history with a different action that I could and should have taken, that would have made no difference whatever in the outcome, but that would have made me feel ever so much better for its primal rightness. You see, I saw that

continued

WHEWELL'S RESTRICTED MEANING OF CONSILIENCE, AS PROPERLY USED BY WILSON

William Whewell (1794–1866) resides among the substantial group of Victorian intellectuals whose interesting lives and deep commitments to learning and new forms of knowledge give them a wider claim upon our memory than conventions for historical fame have granted due to their basically conservative outlook upon the world, or their failure either to make a signal discovery or to attach their name, if only fortuitously, to a memorable concept or an important place (Maxwell's Demon, or the Battle of the Plains of Abraham, named for a local farmer, not the patriarch). The problem for poor Whewell has even become exacerbated by his apparently unpronounceable name. Why try to resurrect a fellow whose moniker you may botch so badly that the few cognoscenti will guffaw during your paper at the next professional meeting? (In this case try, roughly, "you-ull" or "hew-ull" with sufficiently strong emphasis upon the first syllable, and little more than a gulp upon the second, so that the name almost, but not quite, becomes a single syllable—and you'll be close, or so my Oxbridge anglophone buddies tell me.)

Like many conservative Anglican intellectuals, Whewell took theological orders (becoming the Reverend William), but spent his entire career as a university man, in the most proper place of Trinity College, Cambridge, where he

---

young man with the cup of water and I realized what he was about to do. I thought about standing up and just knocking the cup out of his hand, but the incident ended in a second, and I just didn't move quickly enough. Oh, if I had followed my deeply repressed instinct, someone else—probably more people—would have thrown more water on more participants. Maybe a few punches might have been thrown. But so what? The "brave" protestors were poseurs. They presented no threat. Now, I am the very antithesis of a violent man, even for such a largely symbolic action. Yet I long for another shot at that moment, so I could act faster and knock that little cup of water right back into that idiot's face. (Incidentally, folks, stories of this type always "grow" with time. Most reports speak about big pitchers of ice-cold water. Many accounts have even stated that the demonstrators poured blood all over Wilson. It doesn't make a particle of difference because the incident was maximally ugly merely for happening at a scientific meeting at all—but I was there, and the offending item was a small cup of water, thrown by a smaller-minded poseur.)

first served as professor of mineralogy (1828–1832) and a quite competent scientist who, among other achievements, befriended a student named Charles Darwin. He then, beginning to show his range, became professor of moral philosophy (1838–1855); and subsequently, making an almost inevitable plunge into the administrative world, served as Master—their word for boss—of Trinity College (1841 to his death in 1866) and as vice-chancellor of the entire university for a term in 1842 (also meaning boss, because English universities have titular, usually royal, chancellors who do little [by tradition and expectation] but lend their names and preside at commencements and a few other ceremonies, while vice-chancellors do the actual work of an American college president).

Most interesting for our discussion here, Whewell made his signal contributions to science despite his strong religious commitments and titles. (I shouldn't say "despite," for science and religion, as I have argued many times before, do not persist in battle, and many Anglican clergy, in particular, became distinguished scientists, if only because ecclesiastically sponsored study represented one of the few available paths to extensive education for a man like Whewell, who came from an "ordinary" social background. His father was a joiner—and I mean a woodworker, not a glad-hander.) Whewell began his career as a more conventional empirical scientist, doing respectable work in his initial field of mineralogy. As a footnote to history, Whewell actually coined the term *scientist* in 1834, interestingly in a review of a book by the most prominent woman writer on science at the time, Mary Somerville, whom Whewell admired. The word *science* itself enjoyed an ancient pedigree, but in a broader sense to designate any form of knowledge, or *scientia* in Latin—as in my introductory quote from Dryden on page 12. But, for no clear reason, a general name for practitioners of the enterprise had never emerged, a fact that bothered the British Association for the Advancement of Science, which held its first meeting in York in 1831, followed by Oxford in 1832, and then at Whewell's digs in Cambridge in 1833, where the issue received extensive airing in Whewell's presence, leading the noted polymath to do something, ultimately with success, about this odd situation. Thus I should retract part of my previous statement that Whewell's name disappeared from history for lack of a signal achievement among the motley sources of conventional immortality, from gruesome murder to grand discovery. If any of my colleagues can identify Whewell's name at all, the reason will probably reside in vague memory of his paternity for the word *scientist*.

But perhaps Whewell's most distinctive and interesting work lay in his groundbreaking efforts in the history and philosophy of science. Others, including such luminaries as Kant and Voltaire, had treated these subjects before, but in a different and far more explicitly didactic and selective manner. (Whewell, of course, maintained theoretical preferences as well, but one senses his distinctive aim to stress documentation over pure exemplification, and at least to attempt fair coverage rather than directly selective advocacy for a definite point of view.) Whewell added extensive and reasonably balanced descriptions of the history of developing scientific notions to more conventional and selective analyses of right and wrong ways to discover the laws of nature and nature's God. H. Floris Cohen, who holds no special brief of admiration for Whewell, writes in *The Scientific Revolution: A Historiographical Inquiry* (University of Chicago Press, 1994): "What turned Whewell into the man who may rightly be considered the 'father,' or perhaps more fittingly the 'grandfather,' of the historiography of science was his belief that, in order to define in any precise way what these patterns of scientific advance are, one must turn to history."

So Whewell, who could never be accused of sloth, published his three-volume, 1,595-page treatise on the *History of the Inductive Sciences* in 1837, followed three years later by a further two volumes of 1,387 pages, titled *The Philosophy of the Inductive Sciences, Founded Upon Their History.* Whewell explicitly focuses, in both immense books, on the empirical sciences that build conclusions, infer general laws, and devise theories by accumulated and repeated observations and experiments with actual phenomena of nature, rather than by posing abstract mathematical models deduced from the first principles.

In other words, Whewell wished to understand and analyze the process of induction, or movement from repeated observations to general conclusion— the key and definitive activity of successful modern science, in his view— rather than the stronger emphases upon deduction, or logical inference of nature's probable order from more-general principles (perhaps only later tested empirically), as favored by premodern students of the material world. He felt that the strength and power of induction had not been adequately documented, even though Bacon himself, from the earliest seventeenth century, had specified induction as the light and way to modernity in science. Thus Whewell decided to write his two great and sequential treatises on the development and progress of inductive studies about the natural world—first treating the history of scientific progress in three great volumes, and then pulling this material together to explicate, in the subsequent philosophical treatise of

1840, the general powers and pitfalls of induction as the hallmark of scientific advance.

Whewell's definition of consilience occupies much of chapter 5 ("Characteristics of Scientific Induction") and 6 ("Of the Logic of Induction") of the 1840 treatise. He begins by stating that induction can yield general conclusions in two modes, the second more powerful than the first. He names the first of these "colligation of facts," defined as repeated observation, leading eventually to correct prediction, "of facts of the same kind" (page 230)—as, for example, when we decide that water, unlike most fluids, expands when it freezes, because we have, on twenty occasions, filled a crack in a rock with water, allowed the water to freeze, and then noted each time that the resulting ice both exceeds the original water in volume and also splits the rock in two. Then, just to be sure, we even predicted, and then affirmed, that the same result would occur on the twenty-first and twenty-second occasions as well.

But, Whewell then adds, such colligation remains unexpansive (whatever the ice does literally!) in that, by simply repeating a set of identical circumstances over and over, and reaching a generality by induction from invariant results, we only learn a general something about a limited set of objects. We also need to recognize an inductive method that expands outward beyond repeated observation of the same set of occurrences. Hence, Whewell recognizes a second, more powerful, mode of observational inference that he terms "consilience of inductions." Here we face a very different circumstance, frequently encountered in the natural sciences, and bearing the fascinatingly conjoined (and superficially opposite but actually reinforcing) properties of extreme frustration and great potential fruitfulness.

Instead of twenty observations on the same cracking of rocks by ice—a bit boring, but at least leading to a clear prediction and explanation—we now face twenty entirely distinct and apparently disparate observations about a set of objects. But this bundle of facts looks like a total mess. Each may be true and interesting, but each has no apparent bearing upon any other; no thread unites these observations into any commonality. We may have a favored hypothesis about one of these events, but what can we do with the others, for none of these additional events speaks to this hypothesis in any evident way at all?

But then we achieve the great insight for which Whewell invented the beautifully appropriate term *consilience*. We recognize that each and every one of these apparently disparate facts can, after all, be made to cohere—but (and now we state the distinctive feature of consilience), they can so join in one, and absolutely only one, possible way: that is, as consequences of the sole coordinating theory or explanation that could, in principle, bring them thus together into a single, simple, and elegant structure of explanation. Otherwise the facts only stand as disparate and unrelated items of information, with no coordinating or explanatory power beyond their independent existence.

Now, does this kind of situation prove that the single conceivable coordinating explanation must be a correct theory for explaining the joint existence of this large and otherwise entirely incoherent set of facts? Well, such coordination of unrelated bits, as Whewell realizes, does not constitute a formal deductive proof for the proposed explanation. We cannot even say that the coordination corresponds to our usual notion of induction by simple enumeration (the repetitive observation that Whewell had called colligation). But we certainly get a powerful feeling that if one, and only one, kind of explanation can bring all these exceedingly diverse, yet undoubtedly documented, facts together—and no other cause could possibly account for the conjunction in principle—well, what can we do but conclude that this explanation ought to be treated as probably true, or at least as placing us upon a highly useful path to better understanding? What else can one say? At least we should advance potential truth as our hypothesis and try to challenge the explanation by predicting other (presently unrecorded) facts that this coordinating explanation ought to generate.

But what should such a "jumping together" of disparate facts into a common structure of explanation be called? The English language contained no term for this distinctive and important concept, so Whewell took the usual Latin way out and named the process "consilience of inductions"—from *salire*, to jump, and *con*, together: in other words, the "jumping together" of items that appear to be so separate. (I must confess that I always liked Whewell's word because it uses the same root as a famous Latin proverb, cited by Linnaeus, Leibniz, and Darwin as a keystone for their beliefs about natural processes, a conviction that led me to spend much of my career in opposition through studies of punctuated equilibrium and other modes of potentially rapid change in biology: *Natura non facit saltum* (nature does not proceed by jumps). In any case, Whewell defined both the meaning and superior power of his newly minted consilience of induction in the following two paragraphs (page 230):

> The evidence in favor of our induction is of a much higher and
> more forcible character when it enables us to explain and deter-

mine cases of a *kind different* from those which were contemplated in the formation of our hypothesis. The instances in which this has occurred, indeed, impress us with a conviction that the truth of our hypothesis is certain. No accident could give rise to such an extraordinary coincidence. No false supposition could, after being adjusted to one class of phenomena, so exactly represent a different class, when the agreement was unforeseen and uncontemplated. That rules springing from remote and unconnected quarters should thus leap to the same point, can only arise from *that* being the point where truth resides.

Accordingly the cases in which inductions from classes of facts altogether different have thus *jumped together,* belong only to the best established theories which the history of science contains. And as I shall have occasion to refer to this peculiar feature in their evidence, I will take the liberty of describing it by a particular phrase; and will term it the *Consilience of Inductions.*

Whewell gave two primary examples of successful consilience in the triumphs of his nation's greatest scientific hero: Newton's undulatory theory of light and, especially, his inverse square law and the principle of universal gravitation. Whewell explains how Newton's single principle encompasses all three of Kepler's previously uncoordinated laws (pages 230–31):

> It [consilience of inductions] is exemplified principally in some of the greatest discoveries. Thus it was found by Newton that the doctrine of the attraction of the sun varying according to the inverse square of the distance, which explained Kepler's third law of proportionality of the cubes of the distances to the squares of the periodic times of the planets, explained also his first and second laws of the elliptical motion of each planet; although no connection of these laws had been visible before.

More important, Whewell adds, Newtonian gravitation provided a single, and mathematically precise, explanation for two kinds of motion that couldn't seem more different in form or principle: the linear trajectory of an object (apple or otherwise) falling to the earth's surface, and the basically circular motion of the moon around the earth. Whewell summarizes his examples with a lovely phrase from Herschel (page 232):

> The theory of universal gravitation, and of the undulatory theory of light, are, indeed, full of examples of this Consilience of Inductions. With regard to the latter, it has been asserted by Herschel, that the history of the undulatory theory was a succession of *felicities*. And it is precisely the unexpected coincidences of results drawn from distant parts of the subject which are properly thus described.

Ironically—for, as we shall see, Whewell could not abide the later evolutionary theory of his former protégé Darwin—the establishment of evolution as the unifying principle behind the relationships and history of life provides the most instructive case for consilience in all of science. Indeed, I would argue (see Gould, 1986, and, especially, Gould, 2002, for more details than you will ever want to know) that the *Origin of Species* may achieve its most accurate sound bite of description as the most brilliant example ever constructed for the power and efficacy of consilience as a method of proof in natural history. Darwin could not "see" evolution by direct observation in the large (for any substantial change requires more time than humans have inhabited the earth), and he well understood that numerous cases of small change in observable time (breeds of pigeons or dogs, improvement of crop plants) do not prove that large transformations occurred by a similar natural cause. So Darwin used consilience as his primary method. With his unparalleled knowledge of natural history, and his remarkable skills in synthetic argument (he may not have been a great deductive reasoner, but I can think of no greater master of synthesis in the history of science), Darwin constructed the *Origin of Species* as a brief for evolution by consilience. In short, he argues: I present you, in this book, with thousands of well-attested facts drawn from every subdiscipline of the biological sciences—from the transitory and vestigial teeth of embryonic whales, to transitional forms in the fossil record, to the invariant order of life in geological strata throughout the world, to documented cases of small-scale change in agriculture and domestication, to the use of the same bones for such different functions as a horse's run, a bat's flight, a whale's swim, and my writing of this manuscript, to the observation that faunas of isolated oceanic islands always resemble forms of nearby mainlands, but only include creatures that can survive transport across the waters, et cetera, *ad infinitum,* through thousands of equally firm and disparate facts. Only one conclusion about the causes and changes of life—the genealogical linkage of all forms by evolution—can possibly coordinate all these maximally various items under a common explanation. And that common explanation must, at least provisionally, be granted the favor of probable truth.

Moreover, Darwin explicitly attacks creationism most severely for its failure to forge consilience. Over and over again, Darwin tells us how evolution makes coordinated sense of a set of observations, whereas creationism can only regard each separate item as distinct and wondrous. In one passage—a rare expression of annoyance from such a genial man—Darwin explicitly compares creationism to the useless, and nonconsilient, idea once held by some premodern paleontologists that fossils, even though they look just like animals, must have a separate and unknown origin in the mineral kingdom. To provide the full context, Darwin demonstrates how evolution provides a simple and coordinated explanation for the various forms of striping found in the coats of all horse species—from the permanent and prominent coloration of zebras, to the occasional striping of aberrant horses, to weak bands of color that often appear in hybrids between unstriped species, to bands of color that sometimes form in juveniles but disappear in adult life—whereas creationist accounts, with their central premise of a disconnected origin for each species, offer nothing but empty verbiage about divine preferences for order or propensities to craft common signals as aids for human understanding. Darwin compares this lingering mysticism in creationist arguments with a standard caricature supported by describing the foolish delusions of early paleontologists (Darwin, *Origin of Species,* 1859, page 167):

> To admit this [creationist] view is, as it seems to me, to reject a real for an unreal, or at least for an unknown, cause. It makes the works of God a mere mockery and deception; I would almost as soon believe with the old and ignorant cosmogonists, that fossil shells had never lived, but had been created in stone so as to mock the shells now living on the sea-shore.

Nothing so far discussed about Whewell's concept of consilience bears upon any potential conflict between Ed Wilson and me on the relationship of science to the humanities. Nor can we yet grasp why Wilson revived Whewell's word as the title for his book and epitomized description of his program. The resolution lies in the primary implication that Whewell then drew from his concept of consilience, and also regarded as the most important consequence of the idea beyond the basic formulation itself.

In the next pages of his 1840 text, Whewell turns to the question of distinguishing true from false theories—and suggests a property of consilience as a chief criterion. Because consilient theories synthesize apparent mishmashes of large numbers of complex and independent items under the explanatory rubric of a single causal theory (and because consilience seems to point toward true explanations), an additional virtue of consilience should reside in this advantageous property of simplification itself. Therefore, a primary indication of good and true theories should lie in their capacity to *simplify* by subsumption and to *harmonize* by covering disparate items with a single coordinating explanation:

> We have to notice a distinction which is found to prevail in the progress of true and false theories. In the former class all the additional suppositions *tend to simplicity* and harmony; the new suppositions resolve themselves into the old ones, or at least require only some easy modification of the hypothesis first assumed: the system becomes more coherent as it is further extended. The elements which we require for explaining a new class of facts are already contained in our system. Different members of the theory run together, and we have thus a constant convergence to unity. In false theories, the contrary is the case. The new suppositions are something altogether additional;—not suggested by the original scheme; perhaps difficult to reconcile with it. Every such addition adds to the complexity of the hypothetical system, which at last becomes unmanageable, and is compelled to surrender its place to some simpler explanation.

Whewell never exactly extends the argument to Wilson's intricate marriage of reductionism and consilience, but I don't deny that his words about the progress of unification in scientific theory point in that general direction. Whewell certainly identifies, as a major feature of consilience, the discovery that one class of facts achieves a better explanation under "another class of a different nature." And, given reductionistic traditions that have always dominated modern science, no one could accuse Wilson (or anyone else, for I suspect that Whewell himself would have agreed, though he doesn't say so directly) of misrepresenting the intent of this claim by assuming that, in general, the better "class of a different nature" will lie within a science located nearer to the periphery of Ariadne's labyrinth (in Wilson's interpretation)—that is, a more reduced science based upon properties of smaller constituent parts. For Whewell (page 238) describes "that great characteristic of true theory; namely, that the hypotheses which were assumed to account for one class of facts are found to explain another class of a different nature."

Again, Whewell may never tie his concept of consilience directly to a classic chain of reductionism, but the closing words of his chapter 5 surely argue that the continued application of consilience among theories will generate an overall simplification in the structure of scientific explanation—and that such a process leads toward unity by successive generalization. Thus, Whewell certainly links consilience with unification, as an actual consequence if not a logical necessity, in the overall chronology of a trend in science that he would surely have designated as progress (pages 238–239):

> Two circumstances . . . tend to prove, in a manner which we may term irresistible, the truth of the theories which they characterize:—the *Consilience of Inductions* from different and separate classes of facts;—and the progressive *Simplification of the Theory* as it is extended to new cases. [Note Whewell's own emphases by his capitalizations and italics, here retained.] . . . The Consiliences of our Inductions give rise to a constant Convergence of our Theory towards Simplicity and Unity . . . successive steps by which we gradually ascend in our speculative views to a higher and higher point of generality.

In the next chapter, on the logic of induction, Whewell then introduces a metaphorical image, based on consilience, but even more strongly linked to conventional chains of subsumption inevitably suggested by hierarchies of reductionism from maximally complex sciences of large and messy systems like human societies, right down to minimal and highly mathematical theories about a limited number of basic particles that construct all material reality. Whewell now introduces an explicitly genealogical metaphor linked to trees and tributaries flowing into a main river. He even, in one passage (page 244), speaks of "a Genealogical Tree of scientific nobility." Whewell writes in his basic statement (page 241):

> By this means the streams of knowledge from various classes of facts will constantly run together into a smaller and smaller number of channels; like the confluent rivulets of a great river, coming together from many sources, uniting their ramifications so as to form larger branches, these again uniting in a single trunk. The genealogical tree of each great portion of science, thus formed, will contain all the leading truths of the sciences arranged in their due coordination and subordination.

I do not think that Wilson adequately separates Whewell's special meaning of consilience from the general, and far older, scientific procedure (or philosophy) of reductionism, but I surely won't quibble on this point, if only because Whewell himself conflates the two concepts so frequently in stressing the simplifying and coordinating powers of both processes, albeit in their rather different ways (consilience by affirming a particular theory through its sole ability to coordinate otherwise unconnectable facts, and reductionism through its power to resolve complex phenomena by analysis to constituent parts whose simpler, more regular, or more quantifiable properties provide better sources for explanation).

Still, as we shall see in the next section, Wilson really discusses reductionism when he advocates consilience as his basis for uniting the sciences and humanities—and reductionism cannot validate the argument, whereas *Whewell's own meaning of consilience,* which Wilson uses correctly to analyze explanatory styles in science, *cannot be extended into the humanities* for reasons that Whewell himself emphasized (in other important writings) with a commitment as firm as his belief in the validity of consilience within science. But, for now, I must move on to a critique of reductionism and pull a MacArthur for consilience, to which, I promise, I shall return—for the logic works better in this disjointed sequence, as the inadequacy of consilience as an ultimate defense must follow the outflanking of reductionism as an initial strategy.

THE ULTIMATE INADEQUACY
OF REDUCTIONISM WITHIN
THE SCIENCES

Whatever my personal views about full or ultimate success, only a fool or an enemy of science could possibly deny the extraordinary power and achievements of reductionism since the beginning of the Scientific Revolution. Most of the technological accomplishment, and most of the theoretical success, of science emanates from this basic "instinct" for taking complex materials and concepts, breaking them down into smaller constituent parts, and then analyzing the parts, preferably in experimental and quantitative ways, to determine their regularities and, ultimately, the "laws of nature" underlying their repeated and predictable properties.

Thus, for example, Wilson is entirely right (although he calls the process "consilience" in this passage rather than by its rightful name of "reductionism") that the primary reason for the great success versus comparative failure between two disciplines for study of equally complex systems—the medical and the social sciences—lies in the ability of medicine, and the failure of social science, to achieve reduction to sciences of more-basic constituents, better understood and more easily manipulable. Wilson writes (*Consilience,* 1998, page 198):

> The crucial difference between the two domains is consilience:
> The medical sciences have it and the social sciences do not. Medical
> scientists build upon a coherent foundation of molecular and cell
> biology. They pursue elements of health and illness all the way
> down to the level of biophysical chemistry. The success of their
> individual projects depends on the fidelity of their experimental
> design to fundamental principles, which the researchers endeavor
> to make consistent across all levels of biological organization from
> the whole organism down, step by step, to the molecule.

Wilson's dream, and his avowed purpose in writing *Consilience,* lies in his firm belief, frankly proclaimed as a metaphysical assumption and not as a proven scientific reality, that the chain of reductionism heretofore so successful in stretching from particle physics well into the reaches of biological complexity, now (and for the first time) stands poised to make its boldest move upward—starting with (and fundamentally encouraged by) our startling initial successes in beginning to understand the workings of the human brain, and then moving through the social sciences and eventually, and ultimately, into the traditional humanities of arts, ethics, and even parts of religion. He writes (page 9):

> The belief in the possibility of consilience beyond science and across the great branches of learning is not yet science. It is a metaphysical world view, and a minority one at that, shared by only a few scientists and philosophers. . . . Its best support is no more than an extrapolation of the consistent past success of the natural sciences. Its surest test will be its effectiveness in the social sciences and humanities.

Wilson's own testimony provides a best sense of the verve and purpose behind his grand vision of full unification along a single consilient chain of reduction, directed, at least in large part, by his feeling for the elegance and beauty, not to mention the explanatory power and potential emotional satisfaction, that would attend the success of such boldness. And yet, albeit by unintentional use of language (I assume), Wilson also exposes an undiminished belief in the superiority of science, and a devaluing based on misunderstanding the aims and definitions pursued by other forms of knowledge and inquiry—an assumption that cannot forge the kind of allegiances he presumably hopes to establish with scholars in the humanities. For example, his explicit definition, in the following statement, of philosophy as "the contemplation of the unknown," combined with his desire to convert much of this discipline into science (the fruitful study of the knowable and known), will, I am confident, either annoy or at least amuse most professional philosophers. For these scholars, if I understand their enterprise aright, do not define their task as a mere license to speculate or pontificate about things yet unknown, but hold instead that such non-empirical inquiries as rigorous analysis of rules of logic, the structure and classification of argument, and careful examination of the verbal and ideological bases for how people justify and coordinate their beliefs, all constitute valid subjects of study, capable of growth and insight, but not rooted in the scientist's admittedly powerful procedure of validation by the different criterion of coincidence with the structure of material reality, or explanation by general principles that regulate objects and forces in the physical world. (Of course, philosophers will want to know and study everything that the neurosciences can learn—quite a bit already in fact—about our predispositions *not* to reason logically in many circumstances, or even generally. But the full analysis of the logic and rhetoric of arguments lies largely outside the compass of material factuality, and therefore represents another form of intellectual inquiry, fully compatible with, and offering much insight to, the work of science, without simply becoming a topmost branch on the single consilient tree of science.)

> There has never been a better time for collaboration between scientists and philosophers, especially where they meet in the borderlands between biology, the social sciences, and the humanities. We are approaching a new age of synthesis, when the testing of consilience is the greatest of all intellectual challenges. Philosophy, the contemplation of the unknown, is a shrinking dominion. We have the common goal of turning as much philosophy as possible into science. If the world really works in a way so as to encourage the consilience of knowledge, I believe the enterprises of culture will eventually fall into science, by which I mean the natural sciences, and the humanities, particularly the creative arts. These domains will be the two greatest branches of learning in the twenty-first century. The social sciences will continue to split each of its disciplines, a process rancorously begun, with one part folding into or becoming continuous with biology, the other fusing with the humanities. Its disciplines will continue to exist but in radically altered form. In the process the humanities, ranging from philosophy and history to moral reasoning, comparative religion, and interpretation of the arts, will draw closer to the sciences and partly fuse with them.

Other Wilsonian statements underscore his conviction that the bitterness of past failures and the giddy excitement of present possibilities reside in recent scientific advances, primarily in social theory based on evolutionary biology, and more-conventional reductionistic success in understanding the workings of the human brain. For these reasons, and only now, science can finally resume its reductionistic march by breaching the previous wall against a siege that lasted for centuries: our former inability to reach beyond the mechanical functioning and evolutionary history of complex biological forms, especially as expressed in our frustrating failure to penetrate the workings of the brain—what Darwin had called "the citadel itself," and what Descartes, while admitting a material substrate subject to scientific understanding, also regarded as the seat of the soul (located perhaps in the pineal gland), the nonscientific "better" half of his great duality between mind and matter. Wilson (page 66) specifically locates past failures in a conjunction of traditional habits abetted by our previous inability to identify the physical basis and character of mind:

> No intellectual vision is more important and daunting than that of objective truth based on scientific understanding. Or more venerable. Argued at length in Greek philosophy, it took modern form in the eighteenth-century Enlightenment hope that science would find the laws governing all physical existence. Thus empowered, the savants believed, we could clear away the debris of millennia, including all the myths and false cosmologies that encumber humanity's self-image. The Enlightenment dream faded before the allure of Romanticism; but, even more important, science could not deliver in the domain most crucial to its promise, the physical basis of mind. The two failings worked together in a devastating combination: People are innate romantics, they desperately need myth and dogma, and scientists could not explain why people have this need.

I have already outlined, on pages 200–203, why I doubt that the pure reductionistic program, Wilson's full chain of "consilience" if you will, can work either in fact or in principle—and therefore represents the wrong pathway toward such a worthy goal of integration between the sciences and humanities. I do not believe that past failures resided only in a temporary inability to breach a particularly hard barrier (the human mind) in our upward surge to capture new and ever more complex material for inclusion within the reductionistic program. I do revel in the stunning success of the neurosciences. (Speaking personally and emotionally, I could not possibly be more grateful for what we have learned about the genetic components of serious mental disabilities, including the autism of my older son—both for what this knowledge suggests as aid in practical terms, but even more for the emotional and moral liberation thus provided to loving parents previously blamed by formerly canonical psychobabble for bringing on a condition of such pervasive seriousness by some slight, albeit unintentional, suboptimality in parenting.) To such knowledge and liberation, I can only say: give me more and more, quicker and quicker.

Reductionism has enjoyed centuries of triumph, and will continue to fill encyclopedias of additional success. God bless. But just as a sequence in size from mite to *Ultrasaurus* does not imply infinite (or even very distant) further extrapolation, reductionism may not, despite its triumphs in a large domain of appropriate places, be universally extendable as an optimal path to complete scientific understanding. I argued before (see pages 201–203) that two properties of complex systems may deny any dominant status to reductionism, even within the scientific subjects of its evident potential validity; and that, in any case, a third property precludes the incorporation of the humanities into a single consilient chain, whatever the putative success of reductionism in explaining scientific complexity. I shall treat the first, or scientific, reasons here, and save my argument about the humanities for the next section, where my case wins support from Whewell's own negative views about the potential extension of his concept of consilience beyond the natural sciences.

I have already praised Wilson for abjuring the first traditional injury of reductionism—the "inbred" tendency of trained scientists to read its hierarchy of subsumption as a statement about relative worth or "maturity" of the various disciplines—in particular, to praise the charms and colors of particle physics while damning the dismal science of economics. Wilson, as an evolutionary biologist (like me) who works near the disregarded end of this chain, recognizes the chief fallacy of this argument—mere rhetoric for silly personal advantage, usable in either direction (and therefore best avoided at both ends). After all, I might praise particle physics as best because all the other sciences derive from more-complex arrangements of the same particles subject to invariant laws discovered within this most basic of all domains. Yet I might, on the other hand, choose to praise evolutionary biology as the highest science because its levels of complexity, including history and interaction, must apply all the explanatory principles learned from all "lower" sciences on the reductionist chain—so this profession must rank as highest for encompassing most. But then, maybe I should just shut up because such silly posturing reminds me of too many arguments I had at ages nine to eleven on the schoolyard of P.S. 26,

Queens. And I'll be damned if a single one of those altercations ever resulted in anything more positive than my opponent's (very occasional, for I was short, and a less-than-ninety-seven-pound weakling) bloody nose.

I don't wish to belabor one of the most thoroughly adjudicated and widely discussed issues of intellectual life. I will therefore, more as a placeholder than as a claim for adequate coverage (and also, in large part, because the next and rather different argument against incorporating the humanities holds more importance in this general brief), restrict myself to reiterating the two crucial claims generally advanced against the full efficacy of reductionism *within* science. I shall then apply these arguments to a single case, the best recent example in public consciousness, the "deciphering" of the human genome.

1. *Emergence.* This debate has become freighted with all the usual academic impedimenta: confusion in prose and, especially, enormous differences in weights and definitions, ranging from the purely and narrowly technical (my usage here, as I shall explain) to the very broadly religious, with emergence extended and misapplied to unresolvable debates about God's being, meaning, and existence. The basic logical or philosophical question, however, can be posed quite simply. Reductionism works by breaking down complex structures and processes into component parts, and then ultimately explaining the complexity as a consequence of properties and laws regulating the parts.

Now, and obviously, just knowing the properties of each part as a separate entity (and all the laws regulating its form and action as well) won't give you a full explanation of the higher level in terms of these lower-level parts because, in constructing the higher-level item, these parts combine and interact. Thus one must also include these interactions as essential aspects of an adequate higher-level explanation. How, then, can reductionism work if interactions among lower-level parts must figure prominently in any higher-level explanation?

In such cases (effectively including almost any higher-level phenomenon), reductionism still suffices if the interactions can be fully understood and predicted from the parts considered separately. That is, if A and B make C, but if C's distinctive properties arise by predictable necessity from properties inherent in A and B considered separately, then the reduction still works. That is, we still only need to know the components and principles of A and B in order to predict the form and properties of C. The interactions of A and B may impart distinctive properties to C (as we do not taste salt in either pure sodium or pure chlorine). But so long as we can predict these distinctive properties from knowing A and B alone (as we can infer the production of table salt from two components of such different appearance), then reductionism applies.

In technical parlance, interactions predictable from the constituent parts alone are called "additive" or "linear." And so long as interactions remain additive, we can achieve full reduction because nothing must be known or observed exclusively and explicitly at the higher level, in and for itself. That is, we can formulate an explanation, and make correct predictions, simply from our knowledge of the components and their linear interactions.

But suppose that the interactions among constituent parts do not simply cumulate to build the higher-level result by addition. Suppose, to choose an abstract example (representing a pervasive phenomenon in complex systems, I would argue), we wish to study the ecological interaction between species A and B. Suppose we can predict, from the properties of A and B considered separately, that A will always win under a definite set of circumstances. Suppose, just to be sure, that we also approach the issue experimentally, put A and B together on the same field, under the same simple conditions with no other species present, and A indeed wins every time. These results look good for reductionism. But now suppose that when we add species C to the mix, A beats B only half of the time, with B prevailing just as frequently. Suppose also that we carry the example further and discover that the relative frequency of victory for A or B depends upon hundreds of different environmental factors that we can vary at will, obtaining complex but distinctive outcomes each time, and for each set of factors. Maybe A usually wins when D is also present, but always loses when E occupies the field as well.

Maybe, to come to the crucial point, all these complex outcomes fall into an interesting order, even leading to fairly precise predictability about A's or B's eventual prevalence. And maybe, after studying the system for years, we recognize that this clear and complex order cannot be inferred from the components considered separately. That is, we can only achieve our repeatable results by mixing the components together and observing their interactions at their own level of totality. Suppose, finally, that we can even formulate general principles about the nature of these interactions, but only at the level of their direct occurrence. Such kinds of interactions among components are called "nonadditive" or "nonlinear"—and many scientists, myself included, believe that complex systems may well be dominated by such nonadditivity, thereby precluding reductionistic explanation in principle.

For if the domination of nonadditive effects requires that we comprehend the regularity of a system by studying its components as they interact all together, and not by isolating each component, and learning more and more until all the interactions can be predicted from the parts, then reductionism fails in principle. Of course, the reductionist will reply that we have taken the easy and unproven way out. The interactions are surely nonadditive, but perhaps we will be able to know and predict the form of this nonadditivity, once we understand the individual components well enough in our complex but basically deterministic world. In principle, the reductionist claim could be right; and thus the general debate persists among scientists. But I strongly suspect that irreducibly nonadditive interactions pervade natural systems, and that the number, strength, and determinitive power of these interactions increase as systems become more complex—hence the common feeling that, whereas molecular physics may explain the properties of simple chemical compounds in classically reductionist terms, the physiology of individual neurons may not generate an adequate theory of memory.

In any case, and to close with a technical (or definitional) point, properties that make their first appearance in a complex system as a consequence of nonadditive interactions among components of the system are called *emergent*—for the obvious reason that they do not appear at any lower level (and have therefore "emerged" or shown their face for the first time at the new level of complexity). In the strongest form of the argument, we may be able to claim the irreducibility of such emergent features "in principle." For if these emergent properties simply do not exist at the lower level, and can't be inferred, as a consequence of their nonadditive character, from knowledge of lower-level components or their interactions at their own level, then these properties have "emerged" at the higher level, and have no standing within any reduced science on the consilient chain. And, finally, if these emergent properties (as they so often do) become central principles of explanation at the higher level, then reductionism has failed, and the higher level must be studied in its own totality if we hope to achieve satisfactory scientific explanation.

Thus, emergence is not a mystical or antiscientific principle, and certainly provides no brief for any kind of assumption or preference that might be called religious (in conventional terms). Emergence is a scientific claim about the physical nature of complex systems. And if emergent principles become more and more important as we mount the scale of complexity in scientific systems, then the reductionistic research program, despite its past triumphs and continuing importance, will fail both as a general claim about the structure of material reality (the hardest version), or as a heuristic proposition about inevitable (or even most fruitful) ways for advancing scientific knowledge (the weaker or methodological version).

2. *Contingency.* Historical uniqueness has always been a bugbear for classically trained scientists. We cannot deny either the existence or the factuality (yes, the Brits whupped the French at Agincourt in 1415, and the Twin Towers fell on September 11), but we also recognize that no general principle could have predicted the details, and that no law of nature demanded this particular "then and there." Unique facts that didn't have to occur, could never have been predicted beforehand (however much we may later explain the outcomes in fine detail), and will never happen exactly again in all their detailed glory make us very uncomfortable indeed. For we must face (and explain) facts as scientists, but this kind of information does not seem to represent science as we usually understand the concept. We can only hope that we don't need to factor such empirical uniqueness into our explanations very frequently, or in any important way.

And often we are rewarded. Quartz is quartz—and predictably formed when four silicon ions surround each oxygen ion to form a tetrahedron, with each vertex shared between two tetrahedra, yielding the formula SiO_2. Our specimen may have formed a billion years ago in Africa, or fifty years ago in a Nevada bomb crater. We can't even imagine a granting of individuality to the gazillions of tetrahedra in each specimen. Who would dream of contrasting George, the oxygen ion from Africa, with Martha, his counterpart from Arizona?

However, and equally obviously, we do care very much that *Tyrannosaurus* lived in the western United States and apparently became extinct when a large extraterrestrial object struck the earth 65.3 million years ago, and that *Homo sapiens* evolved in Africa, spread throughout the world in short order (evolutionarily speaking), and may not survive the next millennium, a mere geological microsecond. The contrast between the quartz and the creatures may be largely factual, but also includes a strong psychological component that we rarely acknowledge with sufficient clarity. Quartz may represent so simple a system that we couldn't separate George from Martha even if we cared, while a *Tyrannosaurus* would attract notice in human society, even on that fabled New York subway where no one recognizes a well-dressed Neanderthal. But, in large part, we also don't generally give much of a damn about the individuality of simple and apparently repeatable systems; what would we gain, either scientifically or socially, if we could pull out that quartz crystal and say to a friend or colleague: "This is George from the African Cambrian"?

Again, I don't wish to belabor an obvious point (about which I have written *ad nauseam,* even by my standards). Think what you may about reductionism as a procedure for explanation in science. Whatever it may do, however it may work, whatever its range as a favored mode of science; unique historical events in highly complex systems happen for "accidental" reasons, and cannot be explained by classical reductionism. (I do not mean that a kingdom can't be lost for want of a horseshoe nail—that is, that we might trace a very complex outcome to a simple initiating trigger. But the trigger itself can only record another contingency, perhaps of a different level or order. We will not explain Agincourt by the physics of the longbow, or September 11 by the neurology of psychopathology in general, not to mention Mr. bin Laden in particular.)

So, if adequate scientific understanding includes the necessary explanation of large numbers of contingent events, then reductionism cannot provide the only light and way. The general principle of ecological pyramids will help me to understand why all ecosystems hold more biomass in prey than predators, but when I want to know why a dinosaur named *Tyrannosaurus* played the role of top carnivore 65 million years ago in Montana, why a collateral descendent group of birds, called phorusrhacids, nudged out mammals for a similar role in Tertiary South America (at least until the Isthmus of Panama arose and jaguars and their kin moved south), why marsupial thylacines served on the island continent of Australia, and why Ko-Ko both cadged a rhyme and an "in joke" to Katisha when he claimed that he "never saw a tiger from the Congo or the Niger"—well, then I am asking particular questions about history: real and explainable facts to be sure, but only resolvable by the narrative methods of historical analysis, and not by the reductionistic techniques of classical science.

The central importance of contingency as a denial of reductionism in the sciences devoted to understanding human evolution, mentality, and social or cultural organization strikes me as one of the most important, yet least understood, principles of our intellectual strivings. I confess that I have been particularly frustrated by this theme, for the point seems evident and significant to me, and yet I have been singularly unsuccessful in conveying either my understanding or my concern, despite many attempts. Perhaps I am simply wrong (the most obvious resolution, I suppose); but perhaps I have just never figured out how to convey the argument well. Or perhaps—my own arrogant suspicion, I admit—we just don't want to hear the claim.

My point is simply this: Ever since the psalmist declared us just a little lower than the angels and crowned us with glory and honor, we have preferred to think of *Homo sapiens* not only as something special (which I surely do not deny), but also as something ordained, necessary, or, at the very least, predictable from some form of general process (a common position, although defended for obviously different reasons, in the long histories of our professions and within the full gamut of our views on human nature and origins, from pure Enlightenment secularism to evangelical special creationism). In terminology that I have often used before, we like to think of ourselves as the apotheosis of a *tendency*, the end result of some predictable generality, rather than as a fortuitous *entity*, a single and fully contingent item of life's history that never had to arise, but fortunately did (at least for the cause of some cognition on the surface of this particular planet, whatever the eventual outcome thereof).

This mistaken view of ourselves as the predictable outcome of a tendency, rather than as a contingent entity, leads us badly astray in many ways far too numerous to mention. But, in the context of this book's brief for the best way to link science with the humanities, our status as a contingent entity holds special salience as a strong argument against Wilson's favored solution of conjunction by reductive consilience. Because we so dearly wish to view ourselves as something general, if not actually ordained, we tend to imbue the universal properties of our species—especially the cognitive aspects that distinguish us from all other creatures—with the predictable characteristics of standard scientific generalities. When philosophers, from Antiquity on, have analyzed our modes of thinking, and when scientists, from the beginning of our inquiries, have tried to understand our modes of being, these scholars have generally assumed that any identified universal must, *ipso facto,* arise from a lawlike principle, finally manifested at the acme of a tendency embodying all the generality of any natural law or necessity of logic. Thus, whatever we do cognitively (and that no other species can accomplish) becomes part of the definition of cognition as a general principle of complex systems. If our most distinctive property of syntax in language displays certain peculiarities throughout our species, then communication in general must so function. If our arts manifest common themes, then universal aesthetics must embody certain rules of color or geometry.

These subtle, almost always unstated (and probably, for the most part, unconscious) assumptions also prompt the interesting consequence—a serious fallacy in my view—of almost inevitably encouraging a belief that the humanities, if they so embody the only known expressions of phenomena that must represent the highest forms of general and natural tendencies, should be incorporated within science, even though these generalities are, unfortunately for science (which seeks experimental replication above all), expressed in only one species, at least in this world. (This misreading, however, also helps to inspire—and here I mute my criticism because the work can be good, and the questions remain fascinating, whatever the psychological fallacy behind some reasons for the asking—much scientific and semiscientific work loosely coordinated around the theme of trying to make or find another, including attempts to teach language to great apes, work in AI or "artificial intelligence," and the search for intelligent life on other worlds.)

But if *Homo sapiens* represents more of a contingent and improbable fact of history than the apotheosis of a predictable tendency, then our peculiarities, even though they be universal *within* our species, remain more within the narrative realm of the sciences of historical contingency than within the traditional, and potentially reductionist, domain of repeated and predictable natural phenomena generated by laws of nature. And in that case, all the distinctive human properties that feed the practices of the humanities—even the factual aspects that can help us to understand why we feel, paint, build, dance, and write as we do—will, as products of a truly peculiar mind (developed only once on this planet), fall largely into the domain of contingency, and largely outside the style of science that might be subject to Wilson's kind of subsumption within the reductionist chain.

In any case, and to generalize the obvious point, contingency tends to "grab" more and more of what science needs to know as we mount the conventional reductionist chain from the most "basic" science of small, relatively simple, and universal constituents, to the most complex studies of large, messy, multifaceted systems full of emergent properties based on complex webs of massively nonadditive interactions. And although science can study contingency just as well as any other factual subject, such understanding must be achieved primarily by the different methods of narrative explanation, and not by pure reductive prediction. So, as a general statement with many potential exceptions, the "higher" we mount, the less we can rely on reductionism for the twinned reasons of (1) ever greater influence of emergent principles, and (2) ever greater accumulation of historical accidentals requiring narrative explanations as contingencies. The "topmost" fields of the humanities, whose potential for incorporation within the reductionist chain expresses Wilson's primary hope and rationale for his book *Consilience,* seem least likely, for both these reasons, to assume a primary place and definition as the most complex factual systems subject to standard analysis by reductionistic science.

To end with a specific example, the structure of the human genome "met the press" on February 12, 2001. (I will grant some coincidental status to the millennial year, but I know that the choice of both Darwin's and Lincoln's birthday—yes, they were born on the same day, not just the same date, in 1809—recorded a smart and conscious decision in our world of media and symbols.) At this briefing, and with full justification, the press, and the public in general, seemed most surprised by the astonishing discovery that our genome only includes some 30,000 genes, whereas the humble laboratory standard, the fruit fly *Drosophila,* holds half as many, and the far more featureless "worm" of equal laboratory fame, the nematode *C. elegans* (looking like little more than a tiny tube with a bit of anatomical complexity at the genitalia, but virtually nowhere else) has 19,000 genes.

Before this announcement, most estimates had ranged from 120,000 to 150,000, with one company even advertising a precise number of 142,634, and offering to sell their information on individual sequences of genes with potential medical (and therefore commercial) value. This number seemed entirely reasonable because, in some evident sense that even I would not dream of denying, the greater "complexity" of humans, even over the most elegant of nematodes, does seem to require a far greater variety of building blocks as architecture for the intricate totality—and, in common parlance and understanding, each gene ultimately codes for a protein, and congeries of proteins make bodies. So how can humans be so complex with only half again as many genes as a worm—and we refer here not even to a respectably large and somewhat complex earthworm (of a different phylum), but rather to that tiny and featureless, nearly invisible, laboratory denizen, blessed only with the fine name of *C. elegans?*

No one knows the answer for sure, but the basic outline seems clear enough. Genes don't make proteins directly. Rather, they replicate themselves, and they serve as templates for the formation of distinctive RNAs, which then, through a complex chain of events, eventually assemble the vast array of proteins needed to construct a complex human body. One key component of the initial assumption has not, and probably cannot, be challenged. We are, admittedly in some partly subjective sense, far more complex than those blasted worms—and this increment in complexity does require far more components as building blocks. The estimate of 120,000 to 150,000 probably falls in "the right ballpark."

But this number cites the diversity of proteins needed to construct our complexity, and each protein does indeed require a distinctive RNA message as architect. So the 120,000 to 150,000 messages exist, and our previous error must be attributed to a false assumption in the most linear form of reductionistic thinking: namely, that each final protein can be traced back to a distinctive gene through a single chain of causation and construction that, in the early days of molecular biology, received the designation of a "central dogma" (thus showing, by the way, that scientists maintain a decent sense of humor and a high capacity for self-mockery, here shown by expressing a putative basic truth in a manner honorably recognized as overly simplified): "DNA makes RNA makes protein," or the concept of one linear chain of causation extending outward from each gene.

We must also admit that a powerful commercial interest backed this simplest idea that each protein records the coding and ultimate action of a single gene. For if a disorder arises from a particular misconstruction in a specific protein, and if we could sequence the gene coding for this protein, then we might learn how to "fix" the gene, correct the protein, and cure the disease. Thus the debate about patenting genes represented no mere academic exercise for a university's moot court, but rather reflected a driving commercial concern of the large, growing, and highly speculative industry of biotechnology.

Now, of course, whatever one might want to say about scientists, we are not, in general, especially stupid. Just as the very phrase "central dogma" recorded our acknowledgment of a recognized oversimplification, no one ever believed that most diseases would be traced to an easily fixable screw-up in a single protein (although some diseases will be so caused and potentially correctable, and these should be pursued with vigor, provided that we don't deceive ourselves about general theory, and go further astray for the majority of others than we succeed for these fortunate few). And no one ever thought that the simplest form of pure reductionism—a bunch of independent genes, each creating a different protein, one for one, and without any emergent properties to gum up the simple pathways—would describe the embryological construction of the human body. But we do follow an operational tendency to begin with the simplest and most workable model, and then to follow this style of research as far as we can—and we do often make the common mistake of slipping into an assumption that initial operational efficacy might equal ultimate material reality.

Wilson acknowledges these points, and even invokes another humorous acronym to stress the same self-mockery as the central dogma. He expresses more enthusiasm than I could ever muster for the practical range of simplest one-for-one cases, but he also recognizes the probable greater complexity for elaborate mental traits of his primary interest:

> Over 1,200 physical and psychological disorders have been tied to single genes. The result is the OGOD principle: One Gene, One Disease. So successful is the OGOD approach that researchers joke about the Disease of the Month reported in scientific journals and mainstream media. . . . Researchers and practicing physicians are especially pleased with the OGOD discoveries, because a single gene mutation invariably has a biochemical signature that can be used to simplify diagnosis. . . . Hope also rises that genetic disease can be corrected with magic-bullet therapy, by which one elegant and noninvasive procedure corrects the biomedical defect and erases the symptoms of the disease. For all its early success, however, the OGOD principle can be profoundly misleading when applied to human behavior. While it is true that a mutation in a single gene often causes a significant change in a trait, it does not at all follow that the gene *determines* the organ or process affected. Typically, many genes contribute to the prescription of each complex biological phenomenon.

So how do 30,000 genes make up to five times as many messages? Obviously, as we knew (but hoped to identify as a rare exception rather than the evident generality), the linear and independent chains of the central dogma bear little relationship to true organic architecture, and each gene must make (or aid in making) far more than one protein, on average. In fact, we have known (and extensively studied) many potential reasons for at least two decades. The original Watson-Crick models did envision the genome as, to cite the common phrase, "beads on a string"—that is, as a linear sequence of genes stacked end to end, one after the other. But, among many other aspects of genetic structure, two properties of genomes especially discredited the simple bead models long ago. First, the vast majority of nucleotides in the genomes of complex organisms don't code for genes at all, and do not seem to "make" anything of importance to bodies (so-called "junk DNA"). Only one percent or so of the human genome accounts for those *circa* 30,000 genes. Second, and more important, genes are not discrete chains of nucleotides, but are built in pieces of coding regions (called *exons*) interspersed with other sequences of nucleotides that do not translate to RNA (called *introns*). In assembling a gene, the introns are snipped out and the exons joined together to make an RNA from the sequence of conjoined exons. Now, if a gene consists of, say, five exons, we can easily envision several mechanisms for making many different proteins from a single gene. Just consider the two most obvious: either combine the exons in different orders, or leave some of the exons out.

How do the two classical arguments against reductionism fare in the new light of this startling fact—30,000 genes to make five times as many messages, rather than a set of independent linear sequences, each moving from one gene to its necessary protein? I claim no proof in any strict mathematical sense, but rather express a general feeling that the more we learn about complex systems, the less we can sustain a belief that classical reductionism might work, and the more we must suspect that emergence and contingency will enter in ever more important ways as we mount the scale of complexity in nature's material reality. If genes are beads on a string, and each makes a definite protein, then maybe a complex body does "compute" linearly from additive combinations of genes and their decoded products. But if the simple structure of a gene doesn't directly tell us either what protein it will make, how many proteins it will encode, or what parts of what other genes might also be involved in the construction of any protein, then how will we reduce a body made of proteins to the "basic" genetic codes of those proteins? For we now must factor in large classes of new interactions at higher levels not seriously considered before: parts of genes with other parts of the same gene, or with parts of other genes; or complete genes with other complete genes; or less certain linkages and interactions between DNA and various forms and classes of RNA. The more interactions one must consider, and the more these interactions involve larger and larger subunits, the greater the potential importance (and almost sure existence) of emergent principles becomes.

As for contingency, what about all the so-called "junk DNA"? Some may exist for truly random reasons unrelated to natural selection (genes that have, for example, become "unemployed" as their protein products disappeared for good evolutionary reasons from an organism's repertoire, and, thus "disabled," began to accumulate random mutations that destroyed potential function, but that would have been selected out, and predictably so, if the gene were still active). Other components of "junk" may have utilities as yet unimagined, and perhaps at emergent levels not yet understood in the embryology of bodies. In an even more basic sense, if one gene makes one protein, then we could perhaps argue that the body needs each protein it manufactures (and therefore each gene as well), and that bodies therefore represent some form of predictable and optimal conformation. But if 30,000 genes make 150,000 messages, then the same generating set could also make many more messages that do not exist. So an entirely new set of questions now emerges, all leading to thoughts of an enhanced role for contingency. Why these particular 150,000? Why not all the others that could be made? Why make them this way and not by other conceivable routes? The answers to several basic questions of this form must lie in historical accident: yes, *that* could have happened, but *this* did. Either makes sense and can be explained. But this alternative happened to prevail, and something irreducibly fascinating attends explainable claims about basic accidents with such potentially enormous consequences.

WHY REDUCTIONISM CANNOT ENCOMPASS (OR EVEN SUFFICIENTLY INCORPORATE) THE HUMANITIES IN PRINCIPLE

In opening his chapter on the mind, Wilson states his case baldly, with all cards fully exposed (page 105): "Belief in this intrinsic unity of knowledge— the reality of the labyrinth—rides ultimately on the hypothesis that every mental process has a physical grounding and is consistent with the natural sciences. The mind is supremely important to the consilience program for a reason both elementary and disturbingly profound: Everything that we know and can ever know about existence is created there."

As a "benchtop materialist" in practical scientific work, and as an agnostic in religious matters, I completely agree with Wilson's key premise that mental processes have physical groundings and, if knowable at all, must be consistent with the natural sciences. But I strongly disagree with the set of inferences that Wilson then attaches as scaffolding to this putative structure of material reality—leading to his argument for a full, and basically linear, unification of knowledge, a "consilience" (in his terminology) operating in a basically reductionist manner and stretching "upward" through the mind and human social organization into the conventional domain of the humanities and other traditionally "nonscientific" subjects, particularly the arts, ethics, and religion. I presented my first major reason for removing this scaffolding in the preceding section—the argument that reductionism will not suffice even within its potentially applicable domain of subjects traditionally assigned to the natural sciences. In this concluding section, I will develop my second major claim for removal: the argument that, by the logic of its enterprise and the nature of its fundamental questions, the concerns of traditional subjects in humanities (and also in ethics and religion) cannot be addressed and resolved by the methods of scientific inquiry, reductionistic or otherwise.

We should begin by reviewing Wilson's basic argument for extension of his reductionistic chain, his program for "consilience" or unification of knowledge, "above" the sciences of maximally complex systems and directly into the humanities. In the first substantive chapter of his book, following a short introduction, Wilson presents the epitome of his argument, and its implied view of both nature and knowledge. But this short text also includes two caveats and admissions that have spelled the failure of this and similar enterprises for centuries (and will, in my view, continue to do so because the fallacies rest upon errors of logic that cannot be rectified in principle, and not upon missing information potentially "out there" for future collection). Wilson states (page 11) that his quest for consilience

> . . . is equivalent to asking whether, in the gathering of disciplines, specialists can ever reach agreement on a common body of abstract principles and evidentiary proof. I think they can. Trust in consilience is the foundation of the natural sciences. For the material world at least, the momentum is overwhelmingly toward conceptual unity. Disciplinary boundaries within the natural sciences are disappearing, to be replaced by shifting hybrid domains in which consilience is implicit. These domains reach across many levels of complexity, from chemical physics and physical chemistry

to molecular genetics, chemical ecology, and ecological genetics. . . . Given that human action comprises events of physical causation, why should the social sciences and humanities be
impervious to consilience with the natural sciences? And how can
they fail to benefit from that alliance?

The first failing stems from Wilson's own restriction and caveat, "For the
material world at least . . ." Even if I agree that in the material world of science, conceptual unification may be both fully attainable and rapidly pro-
gressing, my concurrence does not extrapolate to an assumption that the
traditional nonscientific subjects of the humanities, ethics, and religion will
join the juggernaut of rapidly expanding oneness. Speaking personally, I suspect that no world other than the material can muster any strong claim for
factual existence. That is, I do not deny Wilson's proposition because we disagree about the structure of the cosmos—he rejecting and I accepting other
forms of undeniable and ascertainable reality, whether we call them spiritual,
divine, or merely immaterial. "Reality" in the sense of what science calls factual truth may well exist only in "the material world," and therefore be entirely
subject to some form of unification (a proposition that can't be proven, but
that would tempt me into a large positive flutter, were I a betting man). But
do all intellectual questions, and all scholarly work, necessarily address "reality" in this form? What about the logic of pure mathematics, with no reference to "stuff" out there? What about inquiry into such trenchant and
poignant subjects as "What must I do in order to say, at the end, that I have
lived a good life"? (A very, very different kind of question from the factual and
anthropological "How do most people and most societies define and practice
the elements that they include in their definition of a good life?") The second
question is terribly important, and lies within the potentially consilient
domain of Wilson's material world. But the first question, far more vital to
most people, simply doesn't reside within the scope of inquiry that scientific
methods can address or factual knowledge decide. And if "the examined life,"
"scholarship," "intellectual inquiry," or whatever you wish to name the serious and professional study of such questions (usually awarded to disciplines
of the humanities) do not fall into Wilson's domain of potential consilience
by reduction, then all knowledge cannot be unified in his sense of the word.

The second failing lies in an erroneous inference made within the penul-
timate sentence: "Given that human action comprises events of physical cau-

sation, why should the social sciences and humanities be impervious to consilience with the natural sciences?" As stated above, I fully accept the first clause, but the second clause simply doesn't follow. Just because my *actions* must be physically caused, and therefore scientifically explainable, I may not infer that criteria of validation, and modes of resolution for *all my questions*, necessarily fall under the same rubric thereby. I cannot evade the laws of physics when I fall, or the laws of neurology when I think, but I may well be able to think about vital things that all people need to address, and scholars have usefully studied, but that fall outside the questions one may ask and answer about the material world of factual reality, where science has been so successful, and where all our lives have been so powerfully impacted.

Before carrying this critique any further, I should state two propositions that, I presume, almost any scientist would find congenial, but that remain fully consistent with my position that the humanities cannot be subsumed into a consilient chain, resting on a reductionistic base, with the sciences. First, I accept that factual information in scientific form will be extremely helpful and relevant to the discussion of almost any important question in nonscientific subjects of the humanities, ethics, and religion. Second, I believe that any humanist who would reject this aid, this helpful hand of friendship, is either a narrow pedant or a fool. In fact, leading scholars in the humanities will actively seek out, and struggle to understand, such factual information, and will try to establish and nurture professional collaborations, and in many cases joint publications, with colleagues in the sciences. (I also believe that, reciprocally, scientists have as much to gain in seeking serious communication with scholars in the humanities.) But, with apologies for any apparent carping, useful collaboration between basically different entities with strong common interests does not imply fusion within a basically linear hierarchy of common structure. One need hardly go beyond the human pair bond (and its status as a base for the villages that raise our children) to appreciate both the structure and potential fruitfulness of different roles for common purposes, or nonfusion for proper diffusion.

My preference for foxes and hedgehogs over labyrinths and chains, as central images for relationships between the sciences and humanities, stems from these objections and distinctions. We do, as scholars, embrace a unity of purpose that might be compared with the well-raised child (filled with knowledge, decency, and discernment, all different but all related to the single goal of wisdom, the hedgehog's one truly great thing). But we also recognize that

many irreducibly different routes, corresponding to the fox's plethora of working pathways, lead to this greatest of all goals. No preferred yellow brick road
can bring us to the Emerald City, a mere confection of wizardry in any case;
but (in a metaphor that I have used previously) we can fashion a coat of many
colors, with each patch necessary to make the completed, glorious cloak of
wisdom. Or, to cut the rhetoric of florid metaphor, and revert to minimal
Latin (another trope of humanism in its most ancient and arrogant form, but
at least in a phrase Americans ought to know): *e pluribus unum.*

As I have, several times before in this chapter, addressed the general issue
of why central, definitive, and nonfactual questions in the magisteria of arts
and ethics (two basic domains within the conventional "humanities") cannot,
in principle, be answered by powerful methods in the magisterium of
science—the crucial support for my case that sciences and the humanities cannot reach, and should not seek, Wilson's form of consilience to achieve the
tighter bonding that nearly all intellectuals strongly favor—I will end by critiquing two particular (but broad) examples of Wilson's proposals, one for the
arts, the other for ethics.

For the arts, Wilson pursues reductionistic consilience by asking if we might
locate the general artistic sensibilities of people in evolutionary reasons for their
origin, then encoded in our neural wiring (and therefore still "with us") as broad
preferences or attractions for certain forms or configurations ("epigenetic rules"
in Wilson's terminology, page 249): "The biological origin of the arts is a working hypothesis dependent on the reality of the epigenetic rules and the arche-
types they generate." Success in such an inquiry would lead to what we might
call a "psychology of aesthetics," a factual understanding of why we prefer (or
even just how we perceive) certain basic forms, and how (or even why, in terms
of evolutionary origins) certain kinds of stories evoke certain emotions and feelings in most people. All fascinating; all eminently useful.

But will the arts, especially in their practice, become a higher branch of
the natural sciences, as success in this form of inquiry accrues? I can see an
artist deriving excellent guidance and suggestions from this new knowledge. I
can even imagine the development of some useful theory about the nature of
what we call human "creativity" in general, whether this attribute be expressed
in answering scientific questions or constructing great works of art. But, art,
at least as I understand the enterprise, is about something (or many things)
fundamentally different from understanding the factual basis of human aesthetic feelings and preferences. A complete neurological analysis of the listener

(quite possible, and undoubtedly interesting) will not explain, in any artistic sense that I seek, the ravishing beauty and emotional power that I experience in Handel's three great Old Testament oratorios of tragic figures felled by their own incubi of madness, bad judgment, or rash vows (*Saul, Samson,* and *Jephtha*). And a complete neurological analysis of the composer (unfortunately impossible) will not explain why Handel was a genius and his contemporary, and then equally popular, rival Buononcini a mere journeyman.*

I don't even deny that broad neurological understandings might aid each of these inquiries, but this level of analysis lies too far from the nonscientific and aesthetic concerns that motivate my feelings and interests. I need to add so much contingency, so much discussion of changing norms and preferences in performance and technique rather than factual truth, so much history, so many personal reasons underlying my affinity for particular pieces or personages. To understand why so simple a piece as the "Dead March" from *Saul* can move me to tears,† I need (at a risible minimum of factors that I can consciously formulate) to analyze Handel's remarkable capacity, again and again, to draw emotion (as Bach rarely, if ever, does) from elementary musical devices—one might almost call them tricks—that appear childishly simplistic. I must then factor in the placement of the piece amid more-complex choral work of Act III, at the drama's height and just before David's impassioned lament for Jonathan, his dear friend and Saul's son, killed in the same battle. I must also add all the memories of a wonderful performance, as inter-

---

*But at least their rivalry generated a wonderful phrase from a contemporary wit—a line that Lewis Carroll later embodied in two fat fellows who declaimed a little ditty about a Walrus and a Carpenter, thus giving Handel's opponent a small place (albeit without his name) in later history:

> Some say, that Signor Bononcini
> Compared to Handel's a mere ninny . . .
> Others aver, to him, that Handel
> Is scarcely fit to hold a candle.
> Strange that such dispute should be
> Twixt Tweedledum and Tweedledee

†I do not hold that science ever claimed explanatory access to so idiosyncratic or trivial a thing as why I love a particular piece. Rather, I only use this example to point out that such central artistic concepts as beauty and passion cannot be accessed without awarding prominence to these intrinsically unscientific factors.

preted by some of the world's greatest Handelian singers, in which I had the privilege of participating as a chorister. I must then figure out why this canonical story of a good man felled by his own insanity (also dramatizing the more general social issue of how societies save themselves when leaders go mad) moves me so much more powerfully than equally universal tales felt more strongly by other people. And then, I must not forget the particulars of contingency, where specific memory may evoke powerful feelings—an especially poignant ingredient in this case, for I shall never forget how the "Dead March," in its searing minimalism, sounded out again and again as John F. Kennedy's casket moved through the streets of Washington one day in 1963.

To generalize Wilson's central misconception, a basic principle of any historical science (including evolutionary biology, where Wilson wishes to locate the origin and meaning, hence the consilience with science, of artistic sensibility) recognizes a key fallacy in equating reasons for historical origin with explanations of current utility for the same structure or behavior. Darwin emphasized the particular evolutionary version of this fallacy, now classically expressed in an example that he could not have known: the impossibility of explaining the current aerodynamic optimality of a bird's feather by reasons for its historical origin from a reptilian scale (for the first feathers covered the arms of small running dinosaurs and could not have served any aerodynamic function, although good models for potential thermodynamic benefits in heat retention have been proposed). Nietzsche (in his *Genealogy of Morals*) then explicitly generalized the argument as a central principle in any historical analysis—in showing how the origin of punishment in a primal "will to power" could neither be inferred from, nor help us to explain, its current range of utilities in modern society, from discouragement of criminal tendencies to encouragement of timely payment for debts.

By this principle, Wilson's linkage of unproven theory for the origin of aesthetic preferences to their current application in artistic judgment cannot even hold within the legitimately scientific realm that I called the psychology of aesthetics. But he then takes the further step of moving from a scientific theory about the origin of aesthetic preferences to validating criteria for truth and beauty in art—the logically fallacious transition from a factual claim of science to an intrinsically nonfactual issue that must be adjudicated, and can be passionately and intelligently discussed (if never "resolved" in the classical terms of science), in the different magisterium of the arts.

Interestingly, Wilson starts modestly (page 230) with a statement that I cannot gainsay, and that harmonizes with the central argument of this book, although Wilson does begin to give his preferences away when he speaks of science's "proprietary sense of the future"—a property that I do not deny, by the way—as a clear "one up" over anything the arts may do in this supposedly equal union (page 230):

> Scholars in the humanities should lift the anathema placed on reductionism. Scientists are not conquistadors out to melt the Inca gold. Science is free and the arts are free, and as I argued in the earlier account of mind, the two domains, despite the similarities in their creative spirit, have radically different goals and methods. The key to the exchange between them is . . . reinvigoration of interpretation with the knowledge of science and its proprietary sense of the future. Interpretation is the logical channel of consilient explanation between science and the arts.

Yet, as his argument develops, Wilson begins to claim more and more territory for natural science in resolving questions in the arts. Just three pages beyond this conciliatory statement, Wilson proposes that consonance with the epigenetic rules of human cognitive function may explain "enduring value" in art. Now, if by "enduring value" Wilson only wishes to make a purely empirical (even measurable) claim about how long, and by how many, a work has been treasured, then he may still be treading within the proper magisterium of science. But if he wishes to conflate such factual conformity to epigenetic rules with "enduring value" in the more usual normative sense of aesthetic worth, then I think that he has run aground on the mudbank of a logical divide:

> Works of enduring value are those truest to these origins. It follows that even the greatest works of art might be understood fundamentally with knowledge of the biologically evolved epigenetic rules that guided them.

Finally, Wilson develops his evolutionary speculations on the adaptive advantage offered by art as the emotional basis for incorporating, by natural selection, certain cognitive universals into the epigenetic rules of human nature. Although I remain unattracted by such basically speculative forms of evolutionary argument, I find Wilson's thoughts both plausible and interesting, albeit unsupported at present. But, at this climax in his discussion of the arts, Wilson now takes the illogical plunge by converting these legitimate speculations about factual and evolutionary origins into explicit claims about the meaning of beauty and truth in art. He begins by arguing that our rapidly increasing intelligence assured our survival and domination, but also exacted a great price (page 245):

> This is the picture of the origin of the arts that appears to be emerging. The most distinctive qualities of the human species are extremely high intelligence, language, culture, and reliance on long-term social contracts. In combination they gave early *Homo sapiens* a decisive edge over all competing animal species, but they also exacted a price we continue to pay, composed of the shocking recognition of the self, of the finiteness of personal existence, and of the chaos of the environment.

"The dominating influence that spawned the arts," Wilson then adds (page 245), "was the need to impose order on the confusion caused by intelligence." We couldn't achieve this control by using our immense brains as flexible computers, and therefore had to encode more-specific cognitive norms of adaptive benefit: "The evolving brain, nevertheless, could not convert to general intelligence alone; it could not turn into an all-purpose computer. So in the course of evolution the animal instincts of survival and reproduction were transformed into the epigenetic algorithms of human nature. It was necessary to keep in place these inborn programs for the rapid acquisition of language, sexual conduct, and other processes of mental development. Had the algorithms been erased, the species would have faced extinction."

But these algorithms, or basic rules of human nature, were too few, too sketchy, and too general to maintain the necessary order all by themselves. So they gained expression as art, thus evoking emotions common and powerful enough to imbue the algorithms themselves with sufficient sway over human actions and propensities (page 246):

> Algorithms could be built, but they weren't numerous and precise enough to respond automatically and optimally to every possible event. The arts filled the gap. Early humans invented them in

an attempt to express and control through magic the abundance of the environment, the power of solidarity, and other forces in their lives that mattered most to survival and reproduction.

In a final paragraph, Wilson makes a doubly false transition: first, from this speculative theory about *origins* to a claim about *current and continuing utility* of the arts; second, and more serious, from a claim in the magisterium of science about the emotional utility of art to a definition of truth and beauty in the magisterium of aesthetics. I may admire the boldness and abruptness of the final claim, but words don't boil rice, and facts of nature or cognition cannot establish a consensus about what art should define as the "beautiful," not to mention the "true."

> The arts were the means by which these forces could be ritualized and expressed in a new, simulated reality. They drew consistency from their faithfulness to human nature, to the emotion-guided epigenetic rules—the algorithms—of mental development. They achieved that fidelity by selecting the most evocative words, images, and rhythms, conforming to the emotional guides of the epigenetic rules, making the right moves. The arts still perform this primal function, and in much the same ancient way. Their quality is measured by their humanness, by the precision of their adherence to human nature. To an overwhelming degree that is what we mean when we speak of the true and beautiful in the arts [page 246].

Turning to ethics, Wilson bases his discussion upon a dichotomy of possible positions that, I thought, had been superseded and gently set aside long ago (with some exceptions, as in Clarence Thomas's defense of "natural law" in explaining his legal views in hearings for his appointment to the Supreme Court; although Thomas won by the thinnest of margins, I don't think that this aspect of his testimony aided his cause—not, to make the obvious point, that such issues have any bearing on so basically political a matter). In contrasting positions that he calls "transcendental" and "empirical," Wilson argues that ethics either record human experience and represent our valid distillation of workable rules for human conduct (the "empirical" view, which would then make ethical precepts subject to factual adjudication and potential reduction to the natural sciences), or derive from a "higher" or more general source independent of our lives, and imposed *a priori* by some universal abstraction or divine will. Wilson begins his discussion by stating his support for the empirical alternative (page 260):

> Centuries of debate on the origin of ethics come down to this: Either ethical precepts, such as justice and human rights, are independent of human experience or else they are human inventions. . . . The true answer will eventually be reached by the accumulation of objective evidence. Moral reasoning, I believe, is at every level intrinsically consilient with the natural sciences.

I regard this setting of the argument as strange, or at least peripheral to the major issue in discussing whether (and how) ethics might shake hands with science. I have little doubt that, on factual matters that might be included in the "anthropology of ethics," the empirical position must prevail, whatever evolutionary reconstruction or interpretation we eventually give to the origin and initial meaning of moral precepts. That is (and I hardly know what other position a modern thinker could take, even a conventionally devout person who has never doubted that ethical truth resides in God's proclamations), I assume that if we surveyed the world's cultures and found that certain ethical principles tended to prevail, we could hypothesize that these principles served a useful function in social organization. If we then found any genetic predisposition for behaviors best suited in practicing these principles, we could also specify a biological and evolutionary linkage to the origin of such beliefs.

Indeed, ever since reading David Hume as an undergraduate, trying like hell to prove him wrong (and failing utterly), I have strongly supported the notion that humans must possess some sort of "moral sense" as an aspect of what we call human nature, and as more than merely analogous with other basic attributes of sight, sound, et cetera. Since ethical "truths" are, in principle, unprovable in any sense that science can recognize (Hume's point, if I understand him aright), I don't know how else we could explain the commonality of certain preferences among various cultures, unless we propose their embodiment in something legitimately called a moral sense.

But how can these propositions address what has always been the crucial and heartrending question about ethics: "How *ought* we behave?"—an entirely different matter from "How do most of us act?" The "is" of the *anthro-*

pology of morals (a scientific subject) just doesn't lead me to the "ought" of the *morality of morals* (a nonscientific subject usually placed in the bailiwick of the humanities).

Wilson, of course, knows that reservoirs of ink have been filled with discussion about whether factual matters can be directly translated into normative or ethical judgments—the famous (or infamous) distinction of "is" and "ought," termed "the naturalistic fallacy" by the early-twentieth-century philosopher G. E. Moore, who evidently argued, in devising his name, that such transitions could not be logically accomplished. But Wilson glosses this issue of the ages by simply stating, more or less, that one obviously can make the move from "is" to "ought" (a prerequisite, needless to say, for the success, or even for the existence, of his program for consilience), and that he can't quite see what all the fuss has been about. In advocating this easy bridge between the anthropology of morals and the morality of morals, Wilson defends what he calls the empiricist position (page 262):

> Ethics, in the empiricist view, is conduct favored consistently enough throughout a society to be expressed as a code of principles. It is driven by hereditary predispositions in mental development— the "moral sentiments" of the Enlightenment philosophers—causing broad convergence across cultures, while reaching precise form in each culture according to historical circumstance. The codes, whether judged by outsiders as good or evil, play an important role in determining which cultures flourish, and which will decline. The importance of the empiricist view is its emphasis on objective knowledge. . . . The choice between transcendentalism and empiricism will be the coming century's version of the struggle for men's souls. Moral reasoning will either remain centered in idioms of theology and philosophy, where it is now, or it will shift toward science-based material analysis. Where it settles will depend on which world view is proved correct, or at least which is more widely *perceived* to be correct.

In an even more incisive statement, Wilson defends the subsumption of ethics into the natural sciences, but then falls into the classical, and still disabling, fallacy in his last line (page 273):

For if *ought* is not *is*, what is? To translate *is* into *ought* makes sense if we attend to the objective meaning of ethical precepts. They are very unlikely to be ethereal messages outside humanity awaiting revelation, or independent truths vibrating in a nonmaterial dimension of the mind. They are more likely to be physical products of the brain and culture. From the consilient perspective of the natural sciences, they are no more than principles of the social contract hardened into rules and dictates, the behavioral codes that members of a society fervently wish others to follow and are willing to accept themselves for the common good.

The argument might just work if we could define "the common good"— the goal of ethical behavior, as Wilson seems to grant—in the objective and empirical terms that subsumption of ethics into the natural sciences inevitably requires. For, once one defines "the common good," then empirical inquiry can determine which behaviors may best achieve the stated goals, and whether (and how) societies have established their ethical rules to reach those ends. But how can we define "the common good," the source of all subsequent arguments, in empirical terms that science may study? Frankly, I don't think that we can—and neither did Hume; nor did G. E. Moore; nor have legions of scholars in the humanities (and the sciences too, for that matter) who have struggled with this issue for centuries, and decided that no single holy grail can exist if several separate streams flow with immiscible waters across the common landscape of our search for wisdom.

How could "the common good" be rendered empirically? The effort stumbles and collapses on the problem that spawned such terms as "the naturalistic fallacy." As I have argued before in this book (see page 142), how can empiricism prevail as the ground of ethics if we discover that most societies, at most times, have condoned as righteous (and validated by ethical rules) a wide variety of beliefs and behaviors—including infanticide, xenophobia (sometimes leading all the way to genocide), and domination and differential punishment of various physically "weaker" groups, including women and children—that most of us strongly wish to repudiate today, with the repudiation, moreover, regarded as the very foundation of a better ethical system? Shall we say that most societies have just been empirically wrong during most of human history—and that we now know better, in much the same way that we once defended a geocentric cosmos and then learned that the earth circles the sun?

Then, in an even more troubling question (that, I suspect, will find a positive answer in empirical terms, and far too often to grant us comfort): how can empiricism prevail as a basis for ethics if we then discover that *Homo sapiens* has indeed evolved biological propensities for the very behaviors that we now wish to repudiate and abjure? What can we say, at this plausible point, except that the empirical anthropology of morals led most societies to a set of precepts with evolutionary origins that may once have made good sense in terms of Darwinian survival—whereas most people have subsequently decided that better morality would lead us to precisely opposite behaviors? How, then, can we avoid the conclusion that the morality of morals (the basis for our decision to forswear an aspect of human nature) must be validated on a basis different from the factual reasons that led our ancestors to adopt moral codes now deemed fit only for rejection on ethical grounds?

At this point, one can hardly avoid the question of questions: If factual nature cannot establish the basis of moral truth, where then can we find it? I don't feel excessively evasive or stupid in admitting that I have struggled with this deepest of issues all my conscious life, and although I can summarize the classical positions offered by our best thinkers through history, I have never been able to formulate anything new or better. After all, if David Hume, and others ten times smarter than I could ever be, have similarly struggled and basically failed, I need not berate myself for coming no closer. I only rejoice that the great majority of good and sensible people in this world seem able to reach a basic consensus on a few central precepts embodied in what we call respect, dignity, and reciprocity, a minimal foundation for enough space and freedom to attempt an ethical life. And if most of these principles sound "negative," as in *primum non nocere* (above all, do no harm), or represent what philosophers call hypothetical rather than categorical imperatives (that is, statements like the Golden Rule based on negotiation and reciprocity rather than upon *a priori* absolutes), then I say bravo for the human decency (an aspect, no doubt, of the "moral sense") that allows us to build reasonable lives on such a flexible and minimal foundation.

Finally, although I reject the possibility of deriving moral principles from empirical study of nature and human evolution, I certainly do not view the divide between "is" and "ought" as utterly impermeable in the sense of claiming that facts can have no relevance for moral thought (although I would defend a strictly logical impermeability in terms of direct movement from natural fact to moral precept). Empirical data will enter any serious discussion of moral principles for a set of obvious reasons, with two rather simple and silly examples listed here as mere placeholders for the generality. First, although technically not illogical, we would be pretty damned stupid (and condemned to utter frustration) if we decided to define something as morally blessed and ethically necessary, even though factual nature declared the feat impossible to attain—as if, for example, we declared the ability to throw a baseball two hundred miles per hour as the chief desideratum of human virtue. Second, and not by any means so inane (but still obvious as the most important impact of factual constraint upon moral struggle), we need to know the factual biology of human nature, if only to gain a better understanding of what will be difficult, and to avoid disappointment at the depth of our struggle, when we properly decide to ascribe moral importance to behaviors that are hard to achieve because they run counter to inborn propensities—as (in a plausible Darwinian inference) for certain forms of cooperation that reduce our own salary or noticeablity, but confer no obvious advantages through the attention or respect gained from others for our altruistic actions.

Wilson, however, still seems to feel that if he can specify the historical *origin* of ethics empirically—a genuine possibility that I regard with optimism—he has solved the basic problem of morality and established a basis for the reduction of ethical philosophy to the natural sciences within his grand chain of consilience. He writes, for example (pages 274–75): "If the empiricist world view is correct, *ought* is just shorthand for one kind of factual statement, a word that denotes what society first chose (or was coerced) to do, and then codified. . . . *Ought* is the product of a material process. The solution points the way to an objective grasp of the origin of ethics."

Yes, the *origin* of ethics would then be within our objective grasp. But such questions about historical beginnings lie in the realm of potential empirics by their very formulation. That is, they reside, in my terminology, within the anthropology of morals, not the morality of morals. Moreover, as stated above, a correct empirical understanding of origins can't reveal the nature of current utility in any case, even if our study were confined entirely within the empirical realm. In other words, I agree with Wilson on the evolutionary origin of ethics, but this issue sits on an irrelevant periphery of the great moral debates in the history of scholarship and human life—non-empirical questions about the meaning of existence and the definition of goodness that science can help us to illuminate and usefully constrain, but that must also, and primarily, be addressed within the logics and methods of the magisterium of the humanities.

WHEWELL'S CONCEPT OF DISCIPLINES, AND A BRIEF ON THE CONSILIENCE OF EQUAL REGARD FOR INTRINSIC BUT COMPLEMENTARY DIFFERENCES

In writing his book on the proper relationship between science and the humanities, Wilson brilliantly chose as his title a single word, alluring through the mystery of its unfamiliarity yet sufficiently comforting in an evident intention implicit in the pleasing aural ring, aided by some etymological hints apparent to most of us in segments of the full term: *consilience*. As explained in the preceding section, Wilson resurrected Whewell's term to express the nub of his proposal for full unification by extending the familiar reductionistic model of the physical sciences "all the way up": breaking through a neurological barrier, into the complexities of social structure and finally into the center of the humanities in arts, ethics, and even religion. This form of unification by reduction joins the humanities to the sciences by granting them topmost positions as empirical studies of maximally complex and various systems, but then asking them, as it were, to "trade" the virtues of the penthouse for what many people and institutions have long and fiercely regarded as the most inalienable of respectful attributes: independence. For, to gain the geometric summit, the humanities must submit their distinctive phenomenologies to explanation by reduction to scientific principles regulating the component parts of their maximal complexity.

Wilson designates this process of unification by successive subsumption into more-tractable principles of "lower" sciences with Whewell's long-lost term *consilience,* or the "jumping together" of disparate facts through their coordinated explanation by simpler and more-abstract laws of correct scientific theories. As discussed previously, reductionism and consilience are not synonymous terms, but Whewell did select classically successful cases of reduction (to the general theory of a "lower" science) as his defining examples of consilience, and he did emphasize his hope and expectation that the "messy" phenomenology of the most complex natural systems would eventually be simplified by consilience under fewer, simpler, and more general laws of the basic physical and natural sciences.

But did Whewell share Wilson's most controversial claim and *sine qua non* of his program for the unification of knowledge: the extension of basically empirical, or scientific, forms of explanation beyond the complex systems that all scholars situate within the domain of science into realms of the humanities as traditionally defined? For the forms of inquiry and the character of defining questions in the humanities would seem to debar explanation in empirical terms—a kind of resolution, moreover, that seems precluded by basic logic, not merely unavailable at present because we lack data that, when discovered, might reduce the cares of the humanities into modes of explanation in the sciences? In fact, and with explicit force in all his major writings, Whewell declared the impossibility of such hegemony for the sciences and their empirical methods.

When one understands Whewell's basic intellectual autobiography, and the general beliefs of his time (particularly of conservative Oxbridge divines like Whewell), one can hardly imagine how a scholar of his import and position could have condoned the extension of his own verbal invention to cover a form of putative unification utterly foreign, indeed dangerous, to his central concept of the nature of the universe, and to his religious beliefs as an ordained Anglican priest with traditional theological commitments. (As I've said, Wilson certainly holds the right to extend Whewell's term into an area not only beyond the intentions of the inventor, but directly contrary to his central view about the nature of knowledge. After all, hardly anyone has used the word at all for more than a century, so any conceivable statute of limitations expired long ago!) Still, the irony of this overturning by apparent extension should be recorded and examined—particularly for my narrower purposes (as I must admit), because Whewell's actual concept of consilience, and of the relationship between science and the humanities, closely matches the position advocated in this book.*

---

*Moreover, speaking of ironies, since both Wilson and I unabashedly own Charles Darwin as our personal hero, and since Darwin's argument for evolution represents the most stunningly successful application of consilience to prove a central theorem in science (see pages 211–212), I must note that Whewell himself rejected Darwin's argument in the *Origin of Species,* and regarded the entire subject of evolution as anathema. As an undergraduate at Cambridge, Darwin admired Whewell as a teacher and celebrated scholar of the sciences. Whewell frequently visited his friend and colleague, the botanist and fellow reverend John Henslow. Darwin often encountered Whewell at these gatherings, for Henslow became the

continued

Latin proverbs about caution extend well beyond dogs *(cave canem)* to certain kinds of people: *cave ab homine unius libri*—beware the man of only one book. One may read this warning in two quite different ways, each keenly appropriate in a distinct manner. One should beware the man who can only write one volume because he never had more than one idea. (The hedgehog, at least, developed a terrific concept for his one great thing; the man of one book usually tries to subsume the entire universe into his cranky and idiosyncratic obsession.) But the other meaning cites the perils of scholarship or posthumous renown, not the limitation of authors. Sometimes we remember a person only for one book, or one achievement, and then, falsely equating the surviving icon with the totality of the actual person, miss the larger and different measure of the man. Goethe was a pretty fair biologist and geologist, and Mickey Mantle was the greatest drag bunter (and fastest runner) in baseball.

---

most important mentor of Darwin's career. In the only reference to Whewell in his short *Autobiography,* written late in life for his children and not intended for publication, Darwin recalls the pleasure and instruction he received in Whewell's company, particularly in their frequent strolls home, following one of Henslow's soirées: "Dr. Whewell was one of the older and distinguished men who sometimes visited Henslow, and on several occasions I walked home with him at night."

But their intellectual relationship soured mightily after Darwin published the *Origin of Species.* In a famous incident and anecdote, the powerful Whewell, as Master of Trinity College, even barred Darwin's book from the college library. Darwin's son Francis recounts the story in his three-volume *Life and Letters* of his father. Recalling the note that Whewell sent to Charles Darwin to discuss his initial response to the *Origin,* Francis Darwin states:

> Dr. Whewell wrote (Jan 2, 1860): ". . . I cannot yet at least, become
> a convert. But there is so much of thought and of fact in what you
> have written that it is not to be contradicted without careful selection
> of the ground and manner of the dissent."

In the next sentence, Francis slyly comments upon Whewell's unnatural selection as expressed in the vigor and practical nature of his opposition (and utterly ineffective as well, for such forms of mild "censorship" only serve to pique interest in what might otherwise have been ignored):

> Dr. Whewell dissented in a practical manner for some years, by refus-
> ing to allow a copy of the "Origin of Species" to be placed in the
> Library of Trinity College.

Whewell frequently falls into this second category of our misreading because his reputation now rests largely upon his great work—albeit two books rather than one, and each of several volumes—in the history and philosophy of the inductive sciences. So we view him as the first modernist with joint command of both history and philosophy in the analysis of science. And since his scientific analyses of modes in reasoning, and progress in accomplishment, were so masterful and unprecedented, Whewell descends to us only in this role of his enduring academic success. But the man fried many other fish in his lifetime, some quite skillfully (although I don't intend to push his versatility to the literal and ludicrous image evoked by my last sentence— a picture of a robed Oxbridge don sweating behind the counter at the local fish-and-chips shop, the quintessential place for an Englishman, now usually a Greek or Pakistani, of different social class and accomplishment).

In particular, although Whewell did not serve as an active parish minister, he was an ordained Anglican clergyman, and he took these commitments seriously, writing several books on religious subjects, beyond the one (or two) of modern memory. This simple statement, all by itself and with nothing added in specific analysis, virtually guarantees that Whewell (unless he expunged some radically dissenting religious views from all his conversations, letters, and writings) could not have advocated the extension of his term *consilience* into the humanities (and particularly into ethics and religion) as a description of their forthcoming union by reduction into the natural sciences. Whewell was a masterful analyst *of science*. But he never imagined, as he invented and explicated such lovely terms as "colligation of facts" and "consilience of inductions," that he had formulated a universal basis for the logic and explanation of all human intellectual endeavors, particularly ethics and religion, the different subject of his other day job. If anything, Whewell stated clearly, again and again, that he had carefully analyzed the ways of science largely to show why this enterprise had been so stunningly successful *in its own domain* of factual nature, and why such an apparatus of explanation could not, in principle, regulate the defining subjects of different logical standing in other magisteria.

In the most notable popular example, Whewell authored, in 1833, the first volume of the famous *Bridgewater Treatises,* a set of (eventually) eight books "on the power, wisdom, and goodness of God, as manifested in the Creation." This authorial bonanza and last gasp for conventional British natural theology originated in a large legacy in the last will and testament of the Right Honorable and Reverend Francis Henry, Earl of Bridgewater, who died in February 1829. And how could any of the eminent people invited to compose these volumes possibly refuse, for the bequest underwrote the printing of 1,000 copies for each book and directed that any profits he paid directly to the authors—a wonderfully effective mixture of Christ and Mammon. The authors, including England's most devout geologist, the Reverend William Buckland, and Mr. Roget of thesaurus fame, did their work quickly and, for the most part, as derivative boilerplate in support of an approach to the natural world that had truly enjoyed its powerful day, but had declined rapidly in public approbation—the so-called "argument from design," or the claim that God's nature and attributes may be inferred from the material character of the cosmos. Most of the volumes did not meet with conspicuous critical success, both because the authors could raise little personal enthusiasm for repeating old concepts expressed so many times before, and because the ideas themselves seemed so dated to many intellectuals, including most theologians. (Darwin and members of his circle generally referred to the series, at least in private letters, as the "Bilgewater Treatises.")

Whewell's contribution, published in 1833 as *Astronomy and General Physics Considered with Reference to Natural Theology,* should, one might suppose at first consideration, support Wilson's notion of consilience. After all, if a noted Anglican priest and prominent scholar in the history and philosophy of science chose to participate in a project dedicated to demonstrating the existence and attributes of God from the products of the material world, then mustn't he accept the central premise of consilience in Wilson's extended form—that is, the direct validation of key subjects in the "highest" nonscientific realm (including the very existence of God) by reduction to the scientific study of nature?

But Whewell takes precisely the opposite course, arguing that whereas God's works (the material cosmos, fully subject to scientific analysis and understanding) cannot conflict with God's words (as revealed in scripture or made known in some other manner), the methods of inquiry and the criteria of explanation differ so profoundly between the two enterprises that meaningful union cannot be achieved by subsuming one domain under the other. Rather, we utilize the two domains to our maximal benefit when we recognize the different light that each can shine upon a common quest for deeper understanding of our lives and surroundings in all their complexity and variety. Whewell felt that he had developed such a long, subtle, and complex analysis of scientific methods and procedures, including his naming and explication of such principles as consilience of inductions, not only to codify and explain the extraordinary efficacy of this conceptual apparatus, but also to demonstrate why these procedures could only work for empirical questions in the factual realm of science and the material world, and could not, in principle, regulate scholarly inquiry in the domain of his other day job as minister, and throughout the grand panoply of subjects rooted in ethical or aesthetic issues, and situated in the humanities.

Whewell summarizes this general argument against consilience (in Wilson's extension) beyond the empirical realm of science in the discussion of morality in his *Bridgewater Treatise.* He begins by stating that moral rules can neither flow from nor contradict the mechanics of empirical nature because ethical discourse rests upon another foundation altogether, with validation achieved by different criteria. Whewell would have equated the distinct foundation of moral rules with a fairly conventional form of Christian testimony; I, along with a majority of Western intellectuals in our times, would seek another source, but would fully embrace Whewell's general point that the basis of morality cannot be established by scientific study of how different ethical standards operate in the empirical world of human cultures. (To grasp Whewell's point in the following quotation, one must also recognize that he uses the term *science* in the old sense of "any legitimate form or body of knowledge," from the literal Latin *scientia,* and not in the restricted modern meaning of factual inquiry about material nature):

The world of reason and of morality is a part of the same creation, as the world of matter and of sense. The will of man is swayed by rational motives; its workings are inevitably compared with a rule of action; he has a conscience which speaks of right and wrong. These are laws of man's nature no less than the laws of his material existence, or his animal impulses. Yet what entirely new conceptions do they involve? How incapable of being resolved into, or assimilated to, the results of mere matter, or mere sense! Moral good and evil, merit and demerit, virtue and depravity, if ever they are the subjects of strict science, must belong to a science which views these things, not with reference to time or space, or mechanical causation, not with reference to fluid or ether, nervous irritability or corporeal feeling, but to their own proper modes of conception; with reference to the relations with which it is possible that these notions may be connected, and not to relations suggested by other subjects of a completely extraneous and heterogeneous nature. . . . There can be no wider interval in philosophy than the separation which must exist between the laws of mechanical force and motion, and the laws of free moral action.

Lest one suspect that Whewell took a less nuanced or more extreme position in his popular writing for the *Bridgewater Treatise* than in his technical monographs on inductive science in 1837 and 1840, his insistence on strict separation between scientific and non-empirical (in this case, religious) forms of knowledge and modes of validation achieves special force and clarity at the very end of the last volume of his 1837 treatise on *History of the Inductive Sciences,* where he treats the emerging study of geology. (Whewell also insists that he has stressed this argument of separation for the benefit of both disciplines, an especially important task when the two magisteria reach the same basic conclusion on a particular matter by following their different routes.) Whewell reserves his strongest statements on the irreducibly different character of science and other sources of knowledge for this closing discussion because the profession of geology, then so young, had been enjoying enormous success, in no small part because its leading practitioners (including such divines as Whewell himself) had abjured an earlier speculative tradition that often blurred the boundaries by seeking explicit theological support for geological assertions (see my discussion of a particular late-seventeenth-century case on pages 77–78).

In no other field of science did the temptation appear so strong, and the tradition remain so well established, to jumble together the immiscible observations of an empirical geological record with the supposedly providential proclamations of scripture. Even if those two sources yielded similar conclusions about the history of the earth, Whewell asserts, we must still keep the inquiries rigidly separate—a principle requiring even more forceful assertion when a supposed agreement in results threatens to encourage a false assumption that distinct ways of knowing might represent different aspects of a single right way after all. Whewell even begins his argument by branding his own ecclesiastical subject matter as "extraneous" to genuine scientific inquiry in geology (1837, volume 3, page 584):

Extraneous considerations and extraneous evidence respecting the nature of the beginning of things, must never be allowed to influence our physics or our geology. Our geological dynamics, like our astronomical dynamics, may be inadequate to carry us back to the origin of that state of things, of which it explains the progress: but this deficiency must be supplied, not by adding supernatural to natural geological dynamics, but by accepting, in their proper place, the views supplied by a portion of knowledge of a different character and order. If we include in theology the speculations to which we have recourse for this purpose, we must exclude them from geology.

Then, on the next page, Whewell forcefully rebuts the same argument for unification of all knowledge along a single chain of rising complexity, with all phenomena subject to one style of explanation, that Wilson, 160 years later, would designate by applying Whewell's own word *consilience* to a theory of knowledge explicitly rejected by Whewell. He begins by acknowledging that all forms of truth must be consistent, while continuing to stress his major claim that consistency of results does not imply a unitary path for ways of knowing (page 586): "It may be urged, that all truths must be consistent with all other truths, and that therefore the results of true geology and astronomy cannot be irreconcilable with the statements of true theology. And this universal consistency of truth with itself must be assented to."

But Whewell immediately follows this statement with his attack upon a single consilient chain (in Wilson's sense), and his defense of irreducibly different ways of knowing, as he ridicules the notion that one might smoothly ascend (or descend) from God's government of his universe to the empirical records of geologic change. (Again we must recognize some key differences between Whewell's definitions and our current understanding of the same words in order to grasp his argument. By "government of the world," for example, Whewell refers to the non-empirical domain of God's ways, not to any putatively factual study of economics or human social organization. We must also remember, as noted before, that "science" refers to any body of knowledge, not only to the empirical realm now employing this term.) Although I do not share Whewell's theology, and would therefore have selected a different example, I have never read a better argument against Wilsonian consilience for the unification of knowledge—all the more striking for expressing the central conviction of the man who invented the word *consilience* (intending thereby only to characterize science more accurately, and all the better to emphasize its inherent logical separation from forms of argument and validation in other magisteria of our intellect):

> To expect that we should see clearly how the providential government of the world is consistent with the unvarying laws by which its motions and developments are regulated . . . is to expect that we may ascend from geology and astronomy to the creative and legislative centre, from which proceed earth and stars; and then descend again into the moral and spiritual world, because its source and centre are the same as those of the material creation. . . . One of the advantages of the study of the history and nature of science in which we are now engaged is, that it warns us of the hopeless and presumptuous character of such attempts to understand the government of the world by the aid of science, without throwing any discredit upon the reality of our knowledge. . . . The error of persons who should seek a geological narrative in theological records, would be rather in the search itself than in their interpretations of what they might find.

Thus, finally, if we must reject Wilson's maximal extension of consilience as the proper strategy for "the greatest enterprise of the mind . . . the attempted linkage of the sciences and humanities" (Wilson, *Consilience*, 1998, page 8), what alternative might better fit the logic of our various intellectual pursuits, also winning thereby a greater opportunity for practical success? Wilson revived Whewell's forgotten word, and extended its meaning far beyond the original authorial intention into a scheme that Whewell himself had strongly rejected—for Wilson wishes to incorporate the humanities into the topmost sciences of a single reductionist chain, thereby achieving a "unification of knowledge" (page 7) under an empirical rubric, whereas Whewell regarded the humanities (particularly moral and religious reasoning) as a set of logically and inherently separate ways of knowing. Serious attention to all members of the set may well unify our mental lives by forging a consensus on values and results. However, such a consensus could only emerge from independent contributions, knitted together by serious and generous dialogue among truly different, and equally valid, ways of knowing, each responsible for a swatch on wisdom's quilt, with the swatches abutting and interfingering in gorgeously complex patterns of interaction. The unification cannot occur (as a logical debarment, not just a practical difficulty) by Wilson's strategy of establishing a single efficacious way of knowing for all disciplines, based on the methods and successes of science, and ultimately valuing the "humanities" not for any intrinsic difference from other factual domains, but for a status as the most complex empirical study of all.

So if Wilson's extension of consilience must fail, and if we still favor the word as a potential description of proper joining for science and the humanities (see my confession of long affection and utilization on page 203), why not try a maximally different strategy? Wilson took Whewell's term for a particular method of validating theories in the inductive sciences, and then generalized consilience to the *ne plus ultra* of possible application by suggesting that all intellectual disciplines, including the humanities, might be unified into a single chain of reductionist explanation rooted in the empirical procedures of science. I would suggest, instead, an opposite scheme of generalization that applies the barest bones of the minimal logic for Whewell's concept to both the sciences and humanities, rather than attempting to unite the two professions into one grand sequence, featuring a single mode of explanation.

This last statement, I realize, may sound cryptic, so let me be more explicit: Whewell defined (and confined) the term "consilience of inductions" to sciences *of a particular kind* for a definite and very interesting reason. Remember that consilience literally means the "jumping together" of disparate observations under the only common explanation that could, in principle, render them all as results of a single process or theory—a good indication, though not a proof, Whewell admits, of the theory's probable validity.

But why and where would one want to employ such a cumbersome method of empirical validation? Why look around for gobs of complex and uncoordinated bits, whose factuality we do not doubt, but whose interrelationship has either never occurred to anyone or has been actively denied because the facts themselves seem so miscellaneous? After all, didn't we learn in high school that science proceeds in the much simpler and far more sensible manner of deducing novel consequences from a hypothesis, and then testing those predictions either to affirm the hypothesis or to throw it out (the far more usual outcome)? As with many idealizations, such a method would, as they say, "be loverly"—if and when we can use it, which, in our real and messy world out there, means "rarely." Whewell did not invent consilience in a book on the philosophy of the *inductive* sciences for capricious reasons, but rather for a wonderfully appropriate motive based on the style of science discussed in his particular treatise.

In sciences that treat relatively simple objects of the physical world, where contingency rarely becomes a crucial component of explanations, where issues of emergence surface infrequently, and where prediction based on the mathematics of invariant natural law often serves as a major device for expanding the compass of a theory—in other words, to continue the stereotyping, for conventionally favored sciences at the base of the reductionist chain—this rational and orderly procedure of prediction and test often works splendidly, and in the advertised manner.

But how do scientists usually proceed in disciplines of maximally complex phenomena, where emergent principles may predominate and contingency often reigns? In such cases—and I used the plethora of facts that Darwin explained by consilience in devising his theory of evolution by natural selection as my primary example (see page 211)—scientists often amass volumes of well-documented, complex, apparently unconnected facts in fields of study boasting few general principles or quantified laws to aid our ordering or explanation. Whewell devised his principle of consilience—again invoking the stereotype of the reductionist chain—primarily for such sciences of maximal complexity at the disfavored top of the standard sequence, where fascinating and intricate phenomenology pervades our observational base, and few general principles impose order or grant clear understanding.

In such cases, featuring a plethora of disparate facts and a dearth of established principles, what shall a scientist in search of a theory do? Whewell did not propose his principle of consilience of inductions as a general guide for simplification and unification of knowledge by subsumption down reductionistic chains (as Wilson suggests). Rather, Whewell developed his concept of consilience as a strategy for devising general theories in difficult sciences of complex systems, which tend to be data-rich and theory-poor. (Such theories, if successful, would then indeed impart highly salutary simplicity of explanation to a previously chaotic system of unconnected facts.)

The intellectual beauty of such Whewellian consilience lies largely in the thrill, even the eeriness, of what current fashion calls an "aha!" experience—the sudden conversion of confusion into order, not by systematic, stepwise, deductive sequences of logical extensions from existing hypotheses, followed by predictions and tests, but rather by an immediate insight that we usually cannot reconstruct in our own psyches because the consilience hits us all at once as from the blue, leading us to emote: "Omigod! All those uncoordinated facts that have tortured me for years in their miscellany do cohere after all"— the "jumping together" that Whewell called consilience because one, and only one, theoretical explanation will array the lot into a sensible order (and, in the best cases, also yield a fascinating and iconoclastic theory as well).

If we focus upon the most general features of this true Whewellian con- silience, perhaps we can formulate, admittedly by extension of Whewell's more restricted intention,* a more adequate statement for the proper rela- tionship between science and the humanities. Consider the components of a Whewellian jumping together: masses of independent items, each separate but equal, and each formerly isolated but now united by a theory or concept recording our desire to bring them together into a coherent and mutually rein- forcing system. In such a crafting of unity by consilience, we cannot specify a higher or lower, a chain of command, or a sequence of reduction or sub- sumption. In contrast to Wilson's hierarchical order of unification by assigned status in a logical and vertical series, we have a true "consilience of equal regard," if you will—a group of formerly independent items, each interesting in itself, each representing something different and enlightening, each perhaps even (for the maximal disparity of disciplines in science and the humanities) immiscible in the logical sense that a satisfactory explanation of one cannot be achieved by modes of resolution favored for the other. The sciences and humanities have everything to gain (and nothing to lose) from a consilience that respects the rich, inevitable, and worthy differences, but that also seeks to define the broader properties shared by any creative intellectual activity, but so discouraged and so often forced into invisibility by our senseless (or at least highly contingent) parsing of academic disciplines. These professional divi- sions, perhaps established for good reasons in some initial time and place, became inadaptive long ago as meaningless separations became hardened by claims for superiority, jargon, incomprehension, ordinary pettiness, fights over university parking spaces, and simple lack of adventurous spirit, combined with the greatest natural impediment to any serious intellectual effort: God's

---

*For I must fairly admit that, like Wilson but in an opposite way, I am suggesting an exten- sion of Whewell's term beyond original authorial usage. I would only argue that my pro- posed extension lies closer than Wilson's to the spirit of Whewell's meaning and does not violate his views on the relations among magisteria of knowledge.

unfortunate limitation of the day to a mere twenty-four hours, and our active professional careers to fewer years than threescore and ten.

I too seek a consilience, a "jumping together" of science and the humanities into far greater and more fruitful contact and coherence—but a *consilience of equal regard* that respects the inherent differences, acknowledges the comparable but distinct worthiness,* understands the absolute necessity of both domains to any life deemed intellectually and spiritually "full," and seeks to emphasize and nurture the numerous regions of actual overlap and common concern. Thus I borrowed our national motto for an epigram and chose the oldest story of the fox and hedgehog for an icon. For our richest form of unification emerges when we can agree on a common set of principles and then derive our major strength for their realization from the different excellences of all cooperating components: *e pluribus unum,* or one from many. Let love of learning be the hedgehog's one great activity, with wisdom as the one great goal. And let us compile a list of necessary components even longer than the effective and inherently different strategems of the fox, with science and the humanities as the two great poles of support to raise the common tent of wisdom.

As I thought about the jumping and leaping of joyful consilience, I remembered one of Isaiah's great prophecies (chapter 35) about coming home

---

*As I have emphasized many times in this book, science surely has nothing to fear from the humanities, and the initial scrappiness of nascent science, fighting for its birthright in the late seventeenth century (see Part I), established unfortunate habits that, abetted by our general human tendencies to parochialism and denigration of others, have persisted for several centuries beyond the extinction of any legitimate rationale (as any supportable reason for this side of the conflict disappeared with the triumph of science, surely by the end of the eighteenth century). Given the power (and cost) of modern science, the suspicions of some modern humanists may claim a more reasonable (or at least a more immediate) basis. But this fear about inequality won't wash either. For every supposed advantage of science, a linked and comparable advantage of the humanities may also be cited. In the most obvious example, science can claim a method capable of ascertaining factual truth, whereas ethical debate in the humanities cannot hope to attain the same kind of confidence about "correct" answers. But we live in a world of trade-offs. Yes, science gains the virtue of factual validation. But even though ethical discourse must sacrifice such a *summum bonum,* who could deny that the basic questions about duties of an ethical life are far more important to our meaning and being. So we swap certainty for salience. As I said, we seek a consilience of equal regard for admitted differences weighed in the balance, with neither side found wanting.

(as my mind replayed Handel's setting of the words in *Messiah*)—an appropriate symbol for the realization of our highest mental and moral possibilities through a consilience of equal regard. We shall all be so much better for the release of impediments: "Then the eyes of the blind shall be opened, and the ears of the deaf shall be unstopped: Then shall the lame man *leap* as an hart, and the tongue of the dumb sing." Isaiah then describes the external benefits of our liberation in a consilience of ethical and intellectual aspects of our being: "In the wilderness shall waters break out, and streams in the desert. And the parched ground shall become a pool, and the thirsty land springs of water."

The way of consilience will open to gather the just, but also to redeem the unwise: "And a highway shall be there . . . and it shall be called The way of holiness . . . the wayfaring men, though fools, shall not err therein." Finally, I read the closing verse of this Chapter 35, and heard Brahms's brilliant setting of the words in his *Requiem:* ". . . songs and everlasting joy upon their heads: they shall obtain joy and gladness and sorrow and sighing shall flee away." Job, of course, will not allow us to forget the idealized and unattainable nature of this reverie, but may we not honor our occasionally and transiently obtainable best by praising the intellect and understanding behind any successful attempt—the fox's many paths to the hedgehog's great place, with science and the humanities linked in a consilience of equal regard? So make a joyful noise *all* ye lands. Didn't the greatest of Enlightenment documents include the "pursuit of happiness" among those few rights that we cannot choose to sell for a mess of pottage, for the search remains as unalienable as intellect itself.

EPILOG

A Closing Tale
of Addition to *Adagia*
by Erasure of Erasmus

AS AN ESSAYIST AT HEART, I HAVE LONG BELIEVED THAT THE BEST, indeed the only effective, discussions of deep generalities begin with intriguing little tidbits that catch a person's interest and then lead naturally to a broad issue exemplified thereby. One simply cannot attack "the nature of truth" head-on, in full and abstract generality, without evoking either boredom or anger at authorial arrogance. But I just disobeyed my own precept by ending the main body of this book with an abstract defense for my version of consilience (versus Wilson's opposite account) as a basic model of proper relationships between science and the humanities. Oh yes, I did bring in the tidbits, circulating throughout the text of this book, of our national motto and the tale of the fox and hedgehog—but only as window dressing, not as the essayist's focal source for a good and expanding story.

So let me try again—one parting shot toward the right track. Let's return to the fox and hedgehog, but this time to specific as specific can be—that is, not just to the old proverb, not just to Erasmus's exegesis thereof, not just to

Gesner's epitome of Erasmus's exegesis and to his wonderfully naive wood-block illustrations of the two animals, but to the treatment of Erasmus's exegesis of the fox and the hedgehog in one particular copy of Gesner's book. We also, as promised, now reengage the titular villain of this book, the anonymous censor who followed the dictates of Magister Lelio Medice, under orders from the Holy Roman Catholic Inquisition of the Diocese of Pisa.

Professional intellectuals form a tiny group in proportion to people who commit their professional lives to a variety of contrary efforts. But "in the beginning was the word," and I wouldn't be pessimistic about the power (or at least the stubborn persistence) of our little fraternity and sorority. They can get us if we fall into a variety of obvious traps, but we usually prevail—or at least we don't go away—if we follow the path of *e pluribus unum,* the conjunction of the fox and the hedgehog, the strategy advocated in this book for proper joining in mutual respect and constant conversation, of the sciences and humanities.

What can be more powerful than combining the virtue of a clear goal pursued relentlessly and without compromise (the way of the hedgehog), and the flexibility of a wide range of clever and distinct strategies for getting to the appointed place, so that someone or something manages to get through, whatever the vigilance and resourcefulness of an enemy (the way of the fox). I regard the consilience of equal regard between science and the humanities as a combination of great power for our small world of scholars because such a joining of truly independent entities, always in close and mutually reinforcing contact, and always pursuing a common goal of fostering the ways and means of human intellect, so deftly combines the different strengths of the fox and the hedgehog that we must win (or at least prevail), so long as we don't allow the detractors to break our common resolve and bond. (Wilson's model of consilience by reductionist unification into a single hierarchy not only misconstrues the inherent nature of similarities and differences in these two intellectual ways, but will also preclude the flexibility of joint expertise in fruitful union by glossing the differences in pursuing the chimera of false unification.)

So I end with a little tale of victory for a particular fox and hedgehog, achieved by combining the contrasting strategies of these exemplary creatures—both for the pleasure of a telling a tiny story with a happy ending, and to restate, in closing, the symbol that I chose, in this book, to carry my central argument for proper relationships of difference, and equal regard in close contact, between the sciences and humanities.

If I may start with an analogy, truth holds many undoubted virtues, from making one free to winning admission, in some systems of belief, to snazzy posthumous places. But among the more abstract benefits, truth certainly shows its greatest practical worth—as Richard Nixon and many others found out to their detriment, as they tried to get by on the opposite path—in allowing a person to keep a complex story straight. After all, if you just tell the truth by your honest recall, you may well be wrong by the vagaries and foibles of memory, but at least you will be calling consistent shots directly, whereas lying and fabrication require that you keep all tiny aspects of an increasingly complex fib constantly in mind, lest you flub and fall into inconsistency by simple failure to remember the details of your past whoppers. Censorship runs into the same practical dilemma as lying. So long as the task remains relatively easy and conceptually clear, an expurgator may perform his odious task quite effectively. But as the kinds and forms of expurgation become more numerous and complex, and especially as the rationale for excision becomes less clear and coherent, even the most anal watchdog will eventually slip, and some dreaded light will sneak through into open rebellion.

Well, Magister Lelio Medice's protégé did pretty well, but Gesner's book, at 1,104 pages, could keep a censor very busy for a long time. Remember that he (see page 56) had been charged with the basically silly and indubitably boring task of expunging all the names of Protestants (including the author Gesner himself) and Catholics of less than fully orthodox bent. (Not really enough to keep the fires of interest burning. Now, a book full of explicit guidelines for witches or pictures of the Bishop of Chichester in the pose of his famous limerick . . . now, that would be something else.)

Basically the censor did little more than blot out names, thousands upon thousands of them, usually several per page, and with no interesting change or deletion of content whatever. Erasmus surely ranked as his greatest challenge and bugbear, for Gesner wrote pages about proverbs for each organism, and Erasmus served as the master source for proverbs. Moreover, Gesner, unlike some modern historians, quoted his sources meticulously. I don't know how many thousands of times Erasmus's name received explicit citation in Gesner's book, but the bedeviled censor had to blot the offending letters out each and every time.

How, ultimately, could the poor censor prevail against Gesner's unconscious but joint employment of the strategies of both fox and hedgehog—so many places and ways to insert offending names, thus decreasing the proba-

408 · De Quadrupedibus

Dipſaco in cacumine capitula ſunt echinata ſpinis, Plinius. Chamæleon candidus ſerp̄ n̄
echini modo ſpinas erigens, Idem. Glycyrrhiza & ipſa ſine dubio inter aculeatas eſt, ſol̄
tis, Plinius: ego nihil tale in glycyrrhiza noſtra hactenus deprehendi. Echinopus Ath̄
neſcio quam herbam ſpinoſam ſignificat, quam poëta quidā unā cum ononide nominat, Σ̄
βας συνάγων, ὡς ἀν̄ ἐχινόπος ας κȣ ἀνὰ προκείαν ὄνωνιν Ἀεῖ ϛατρίβων, εὐθεωρ τὴς ἡδίςης μ̄
in grammaticum quendam anxiè diligentem circa ſingulas uoculas, ſolidæ uerò eloquē
ditionis negligentem. Ceras ex omnium arborum ſatorumꝗ floribus apes confingunt, c̄
mice & chenopode: Herbarum hæc genera, Plinius 11.8. malim echinopode: quoniā ch̄
nomen nuſquam inuenio: conuenit autem herbæ ſpinoſæ ab echino factum nomen.

¶ Inſulæ Echinades dictæ ſunt ab Echione quodam, uel à multitudine echinorum, ſ̄
illi, ſiue marini fuerint: Vel quod ſolum earum aſperum & ſpinoſum ſit echinorum inſtar, E
in Dionyſium. Oxiæ inſulæ, quas Homerus Thoàs uocauit, Echinadibus propinquæ ſ̄
inter eas à Strabone collocentur, Hermolaus. Echinæ inſulæ ſunt circa Aetoliam, quib̄
fluuius limum adijcit, Echinades aliàs dictæ, ἔςι τὸ προκὺ κȣ ὀξὺ; uel ab echinorum copia, ū
Iſodoro placet ab Echino uate, Stephanus. Echinades, inſulæ Acarnaniæ iuxta hoſtia Ach̄
in quibus Epei dicti habitãt, Scholia in Iliad.2. Plura quære in Onomaſtico noſtro. Echī
eſt ciuitatis, cuius (uel uiri à quo dicta eſt, ut Etymologus habet) meminit Demoſthenes Ph̄
ta, Suidas. Echinos urbs eſt Acarnaniæ ab Echino condita, quam Rhianus Ἔχον ἀςν uō
Echinûntem, Stephanus. Echinus urbs Theſſaliæ ſic dicta ἀπ̄ Ἐχίνε ἑνὸς τȣ τὼ Σπερχ̄
Varinus: uel ἑνὸς τῶ Σπαρτῶν (lego ασπαρτιατῶν) ᾠνοῦθα οἰκήσαντ᾽, Etymologus, Hanc ē
Phthiotide collocat Ptolemæus, in faucibus Sperchij amnis, Plinius. Echinûntis mentio f̄
ceronem in Arato, Dicitur excelſis errans in collibus amens, Quos tenet Aegeo delɔ̄ ū
Echinus. Echinos Thraciæ urbs ad Pagaſeum ſinum, Pomponius lib.2. Sperchium t̄
liacum ſinum deſinere Ptolemæus ſcribit.

¶ c. Mures alpini totam hyemem in latibulis uſꝗ ad uer erinaceorū inſtar conuoluti d̄
& dormiunt, Ge. Agricola.

¶ e. Sanguine herinacei cum decollatur, æquali oleo mixto, ſi intingatur corpus ū q̄
tis quid ſit, ligatur ab omnibus mulieribus uſꝗ ad menſem, Raſis & Albertus. Oculū
dexter frixus ad pondus unciæ (Raſis neſcio quas ponderū notas hic habet) cum oleo ā
alnulæ) uel ſeminis lini, ſi ponatur in uaſe æris rubri, & collyrij modo inde illinantur oc̄
qui noctu uidere deſyderat, in tenebris condita quælibet iam uiſu diſcernet quàm inter̄

¶ h. Magos qui Zoroaſtren ſectantur, imprimis colere aiunt herinaceum terreſtrē
uerò odiſſe mures aquaticos, Plutarchus in Sympoſiacis lib.4. quæſtione ultima. Idē
Iſide, terreſtres echinos ab his magis bono deo attribui ſcribit, aquaticos autem malo. D̄
ſacrificato ſupra dixi capite primo.

¶ PROVERBIA. Echino aſperior, Ἐχίνε τραχύτερῷ, in hominem intractabilem &̄
moribus dictum, metaphora ſumpta ab echino ſiue terreſtri ſiue marino, **Sufra.** (T̄
nus aſper, Ἄπις ἐχῖνῷ τραχύς, in moroſos & iniucundis moribus quadrat, Echini enim t̄
tum marini undiꝗ ſpinis obſepti ſunt, ut nuſquam impune poſſis attingere. Eſt & hom̄
modi genus cum quibus nulla ratione poſſis agere citra litem. Ariſtoteles in Pace, ο̄
λείον τον τραχὺ ἐχίνον, id eſt, Ex hīto in læuem nunquam mutabis echinum, Eraſmus. Schol̄
ſtophanis aptum huius dicti uſum eſſe oſtendit, cum quis alicui infenſus & aſper, mitis ā
erga ipſum ut fiat perſuaderi non poteſt. Echinus partum differt, Ἐχῖνῷ τὸν τόκον ā
ci ſuetum qui prorogarent quippiam ſuo malo: ueluti qui creditam pecuniam compē
men aliquando reddendam uel maiore cum fœnore. Aiunt echinum terreſtrem ſtimul̄
rari partum, deinde iam aſperiore ac duriore facto fœtu mora temporis, maiore cruciat̄
thor Suidas, **Sufra.** Echinus parturiens cunctatur: uel Echinus partum procraſtinat p̄
in eos qui in perniciem ſuam morarum cauſas innectunt: cuiuſmodi ſunt illi uerſuram̄
Budæus. ¶ Prius duo echini amicitiam ineant, alter è mari, alter è terra, πρὶν δ᾽ δύο ἰ̄
ἔλθωσιν, ὁ μ̄ ἐκ πελάγος, ὁ δ̄ ἐκ χέρσο: de ijs qui moribus ac ſtudijs ſunt inter ſe diſcrepā
ſpes ſit aliquando inter eos neceſſitudinem coituram. Refertur à Suida, **Sufra.** (M̄
uulpes, uerum echinus unū magnum, πολ̄ δίδ᾽ ἀλώπηξ, ἀλλ᾽ ἐχῖνῷ ϛν μέγα, Zenodotus h̄
ex Archilocho citat. Dicitur in aſtutos, & uarijs conſutos dolis. Vel potius ubi ſignific̄
dam unica aſtutia plus efficere, quàm alios diuerſis technis. Nam uulpes multijugis dol̄
aduerſus uenatores, & tamen haud rarò capitur: Herinaceus unica duntaxat arte tut̄
canum morſus. Siquidem ſpinis ſuis ſemet inuoluit in pilæ ſpeciem, ut nulla ex parte mō
di queat, **Sufra.** Ὅτι δ̄ οἱ ἐχῖνοι, λέγω δ̄ κȣ οἱ χερσαῖοι κȣ οἱ θαλατίοις, κȣ ἑαυτ̄ εἰσι φυλ̄
θηρώμας πȣ βαλλόμενοι τὰς ἀκάνθας, ὥσπερ τι χαράκωμα, ἵων ὁ Χῖῷ μαρτυρεῖ οἱ φοινίκης κȣνι, λέγω δ̄
ᾠ τι χέρσω τὰς λίοντῷ ἠρίζ᾽ Ἡ τὰς ἐχῖνς μᾶλλον ἀίυρας τέχνας. Ὡς ἐντ̄ ἀ̄ ἄλλων δνείων ὁτρφ ρ̄
Βιλῷ ἀμφ᾽ ἀκανθαν εἰλίξας ὀίεμας Κῶται, δ᾽ ἀκαίρ τι κȣ θιγεῖν ἀμύχαντ᾽, Hæc ex Athenæi Dȳ
lib. 3. Eraſmus eoſdem uerſus ex Zenodoto recitat, & primo quidem uerſu pro ᾠ legit γ λ̄
do autem pro ἡ legit κȣ, & ita transfert; Leonis artes in ſolo ſanè probo, At magis echini cō

bility of finding every last one (the fox's flexible strategy); and so many sim-ple and mind-numbing repetitions of the same name in the same basic con-text, Erasmus on proverbs, Erasmus on proverbs, Erasmus on proverbs (the hedgehog's one stubborn mode). The poor man just couldn't be perfect against such an onslaught of dull heresy. And guess where he fell?

If we turn to the page on proverbs in Gesner's chapter *De Echino* (on the hedgehog), we find the name of Erasmus, dutifully blotted out four times (see figure 32). But there, right smack in the middle of Gesner's discussion of Erasmus's exegesis of Archilochus's old legend about the fox and the hedge-hog, our man in Pisa finally slipped and let Erasmus's name pass through in the most symbolic spot of all. So let the unblotted name of the great Erasmus of Rotterdam, discussing the old and enigmatic motto of the fox and the hedgehog, represent the necessary victory of our best intellectual (and ethical) inclinations, provided that we stick together in our broad and useful diversity, following the ways of both creatures as we exploit all honorable paths to the hedgehog's one great thing of wisdom.

I ended the main body of this book with a reference to America's most famous Enlightenment statement by the estimable Mr. Jefferson (see page 260). I will now truly finish by citing the even more estimable Mr. Franklin, our greatest Enlightenment hero, in one of the finest English puns ever minted. As he stated for the people of America, and for the thirteen colonies of *e pluribus unum*—and as I say for the wonderful and illuminating differ-ences between the sciences and the humanities, all in the potential service of wisdom's one great goal—we had better hang together, or assuredly we will all hang separately.

Figure 32.

INDEX

Note: Page numbers in *italics* refer to illustrations.

O

P

Q

R